Advancing into Analytics

From Excel to Python and R

George Mount

Beijing · Boston · Farnham · Sebastopol · Tokyo

Advancing into Analytics

by George Mount

Published by O'Reilly Media, Inc., 1005 Gravenstein Highway North, Sebastopol, CA 95472.

O'Reilly books may be purchased for educational, business, or sales promotional use. Online editions are also available for most titles (*http://oreilly.com*). For more information, contact our corporate/institutional sales department: 800-998-9938 or *corporate@oreilly.com*.

Acquisitions Editor: Michelle Smith	**Indexer:** Sam Arnold-Boyd
Development Editor: Corbin Collins	**Interior Designer:** David Futato
Production Editor: Daniel Elfanbaum	**Cover Designer:** Karen Montgomery
Copyeditor: JM Olejarz	**Illustrator:** Kate Dullea
Proofreader: Justin Billing	

April 2021:   First Edition

Revision History for the First Edition

2021-04-15:   First Release
2021-09-17:   Second Release
2022-10-21:   Third Release

See *http://oreilly.com/catalog/errata.csp?isbn=9781492094340* for release details.

978-1-492-09434-0

[LSI]

Table of Contents

Part II. From Excel to R

Part III. From Excel to Python

Preface

You're about to partake in a significant and commendable learning journey that will involve statistics, coding, and more. Before diving in, I'd like to take some time to address my learning objectives for you, how I arrived at this book, and what you should expect.

Learning Objective

By the end of this book, you should be able to *conduct exploratory data analysis and hypothesis testing using a programming language*. Exploring and testing relationships is core to analytics. With the tools and frameworks you'll pick up in this book, you will be well positioned to continue learning more advanced data analysis techniques.

We'll be using Excel, R, and Python because these are powerful tools, and because they make for a seamless learning journey. Few books cover this combination, even though the progression from spreadsheets into programming is common for analysts, myself included.

Prerequisites

To meet these objectives, this book makes some technical and technological assumptions.

Technical Requirements

I am writing this book on a Windows computer with the Office 365 version of Excel for desktop. As long as you have a paid version of Excel 2010 or greater for either Windows or Mac installed on your machine, you should be able to follow along with the majority of the instruction in this book, with some variations, particularly with PivotTables and data visualization.

While Excel offers both free and paid versions online, a paid desktop version is needed to access some of the features covered in this book.

R and Python are both free, open source tools available for all major operating systems. I address how to install them later in the book.

Technological Requirements

This book assumes no prior knowledge of R or Python; that said, it does rely on moderate knowledge of Excel to flatten that learning curve.

The Excel topics you should be familiar with include the following:

- Absolute, relative, and mixed cell references
- Conditional logic and conditional aggregation (IF() statements, SUMIF()/ SUMIFS(), and so forth)
- Combining data sources (VLOOKUP(), INDEX()/MATCH(), and so forth)
- Sorting, filtering, and aggregating data with PivotTables
- Basic plotting (bar charts, line charts, and so forth)

If you would like more practice with these topics before moving on, I suggest *Excel 2019 Bible* by Michael Alexander et al. (Wiley).

How I Got Here

Like many in our field, my route to analytics was circuitous. In school, mathematics became a subject I actively avoided; too much of it seemed entirely theoretical. I did have some coursework in statistics and econometrics that caught my interest. It was a breath of fresh air to *apply* mathematics to some concrete end.

This exposure to statistics was admittedly scant. I attended a liberal arts college, where I picked up solid writing and thinking skills, but few quantitative ones. When I got to my first full-time job, I was floored by the depth and breadth of the data I was entrusted with managing. Much of this data lived in spreadsheets and was hard to get much value out of without intense cleaning and preparation.

Some of this "data wrangling" is to be expected; the *New York Times* has reported that data scientists spend 50% to 80% of their time preparing data for analysis (*https:// oreil.ly/THah7*). But I wondered if there were more efficient ways to clean, manage, and store data. In particular, I wanted to do this so I could spend more time *analyzing* the data. After all, I always found statistical analysis somewhat palatable—manual and error-prone spreadsheet data preparation, not so much.

Because I enjoyed writing (thank you, liberal arts degree), I started blogging about tips I picked up in Excel. Through good grace and hard work, the blog gained traction, and I attribute much of my professional success to it. You are welcome to stop by at *stringfestanalytics.com*; I still post regularly on Excel and analytics more generally.

As I began to learn more about Excel, my interest spread to other analytics tools and techniques. By this point, the open source programming languages R and Python had gained significant popularity in the data world. But while I made my way through grasping these languages, I felt unnecessary friction in the learning path.

"Excel Bad, Coding Good"

I noticed that for Excel users, most R or Python training sounded a lot like this:

> All along, you've been using Excel when you really should have been programming. Look at all these problems Excel has caused! Time to kick the habit entirely.

That's the wrong attitude to take for a couple of reasons:

It's not accurate
The choice between coding and spreadsheets is often framed like a sort of struggle between good and evil. In reality, it's better to think of these as *complementary* tools rather than substitutes. Spreadsheets have their place in analytics; so does programming. Learning and using one does not negate the other. Chapter 5 discusses this relationship.

It's a poor instructional approach
Excel users intuitively understand how to work with data: they can sort, filter, group, and join it. They know which arrangements make for easy analysis, and which mean lots of cleanup. This is a wealth of knowledge to build on. Good instruction will use it to bridge the gap between spreadsheets and coding. Unfortunately, most instruction instead burns the bridge out of contempt.

Research indicates that relating what you've learned to what you already know is powerful. As put by Peter C. Brown et al. in *Make It Stick: The Science of Successful Learning* (Belknap Press):

> The more you can explain about the way your new learning relates to your prior knowledge, the stronger your grasp of the new learning will be, and the more connections you create that will help you remember it later.

As an Excel user, it can be hard to relate new ideas to what you already know when you're (wrongly) told that what you already know is garbage. This book takes a different approach, building on your prior knowledge of spreadsheets so that you'll have a clear framework in mind as we move into R and Python.

Both spreadsheets *and* programming languages are valuable analytics tools; there's no need to abandon Excel once you've picked up R and Python.

The Instructional Benefits of Excel

In fact, Excel is a uniquely fantastic analytics teaching tool:

It reduces cognitive overhead

Cognitive overhead is the number of logical connections or jumps needed to understand something. Often an analytics learning journey looks like this:

1. Learn a brand-new technique.

2. Learn how to implement the brand-new technique using brand-new *coding* techniques.

3. Progress to more advanced techniques, without ever having felt really comfortable with the basics.

It's hard enough to learn the conceptual foundations of analytics. To learn this while *also* learning how to code invites sky-high cognitive overhead. For reasons I'll discuss, there is great virtue in practicing analytics via coding. But it's better to isolate these skill sets while mastering them.

It's a visual calculator

The first mass-market offering of a spreadsheet was called VisiCalc—literally, a visual calculator. This name points to one of the application's most important selling points. Especially to beginners, programming languages can resemble a "black box"—type the magic words, hit "run," and presto: the results. Chances are the program got it right, but it can be hard for a newbie to pop open the hood and see why (or, perhaps more critically, why not).

By contrast, Excel lets you watch an analysis take shape each step of the way. It lets you calculate and recalculate visually. Rather than just taking my (or a coding language's) word for it, you'll build demonstrations in Excel to visualize key analytics concepts.

Excel provides the opportunity to learn the fundamentals of data analytics without the need to learn a new programming language at the same time. This greatly reduces cognitive overhead.

Book Overview

Now that you understand the spirit of the book and what I hope for you to achieve, let's review its structure.

Part I, "Foundations of Analytics in Excel"
Analytics stands on the shoulders of statistics. In this part, you will learn how to explore and test relationships between variables using Excel. You'll also use Excel to build compelling demonstrations of some of the most important concepts in statistics and analytics. This grounding in statistical theory and framework for conducting analysis will put you on solid footing for data programming.

Part II, "From Excel to R"
Now that you are fluent in the fundamentals of data analysis, it's time to pick up a coding language or two. We'll start with R, an open source language built especially for statistical analysis. You will see how to cleanly transfer what you've learned about working with data from Excel into R. I conclude the section with an end-to-end capstone exercise in R.

Part III, "From Excel to Python"
Python is another open source language worth learning for analytics. In the same spirit as Part II, you'll learn how to pivot your Excel data chops into this language and conduct a complete data analysis.

End-of-Chapter Exercises

When I read books, I tend to skip over the exercises at the end of the chapter because I feel keeping the momentum of my reading is more valuable. *Don't be like me!*

I provide opportunity at the end of most chapters to practice what you've learned. You can find the solutions to these drills in the *exercise-solutions* folder (*https://oreil.ly/KVrIn*) of the accompanying repository, where you'll see them in a file named for each respective chapter. Complete these drills, then compare your answers to the solutions. You'll be increasing your comprehension of the material while at the same time providing a good example for me.

 Learning is best done actively; without putting what you've read into immediate practice, you're likely to forget it.

This Is Not a Laundry List

One thing I love about analytics is there are almost always multiple ways to do the same thing. It's likely I'll demonstrate how to do something one way when you are familiar with another.

My focus for this book is on using Excel as a teaching tool for analytics and helping readers transfer this knowledge to R and Python. Were I to make this a brain dump of all the ways to complete a given data-cleaning or analysis task, the book would lose its focus around this particular objective.

You may prefer alternative ways of doing something; I may even agree with you that, given different circumstances, there is a better approach. However, given the circumstances *of this book* and its objectives, I've decided to cover certain techniques and exclude others. Doing otherwise would risk turning the book into a bland how-to manual rather than a pointed guide for advancing into analytics.

Don't Panic

As an author, I hope you find me easygoing and approachable. I do, however, have one rule for this book: *don't panic!* There is an admittedly steep learning curve at play here, since you'll be exploring not just probability and statistics but *two* programming languages. This book will introduce you to concepts from statistics, computer science, and more. They may initially be jarring, but you'll begin to internalize them over time. Allow yourself to learn by trial and error.

I thoroughly believe that with the knowledge you possess about Excel, this is an achievable order for one book. There may be moments of frustration and impostor syndrome; it happens to all of us. Don't let these moments overshadow the real progress you'll make here.

Are you ready? I'll see you over in Chapter 1.

Conventions Used in This Book

The following typographical conventions are used in this book:

Italic
> Indicates new terms, URLs, email addresses, filenames, file extensions, and dataset variables.

`Constant width`
> Used for program listings, as well as within paragraphs to refer to program elements such as code variable or function names, databases, data types, environment variables, statements, and keywords.

This element signifies a tip or suggestion.

This element signifies a general note.

This element indicates a warning or caution.

Using Code Examples

Supplemental material (code examples, exercises, and so forth) is available for download at *https://github.com/stringfestdata/advancing-into-analytics-book*.

You can download and decompress a copy of the folder on your computer or, if you are familiar with GitHub, clone it. This repository contains completed copies of the scripts and workbooks for each chapter in the main folder. All datasets needed to follow along with this book are located in a separate subfolder of the *datasets* folder, along with notes about its source and steps taken to gather and clean it. Rather than operate directly on any of these Excel workbooks, I suggest you make copies, as manipulating the source files may affect later steps. All solutions for the end-of-chapter exercises can be found in the *exercise-solutions* folder.

If you have a technical question or a problem using the code examples, please email *bookquestions@oreilly.com*.

This book is here to help you get your job done. In general, if example code is offered with this book, you may use it in your programs and documentation. You do not need to contact us for permission unless you're reproducing a significant portion of the code. For example, writing a program that uses several chunks of code from this book does not require permission. Selling or distributing examples from O'Reilly books does require permission. Answering a question by citing this book and quoting example code does not require permission. Incorporating a significant amount of example code from this book into your product's documentation does require permission.

We appreciate, but generally do not require, attribution. An attribution usually includes the title, author, publisher, and ISBN. For example: "*Advancing into Analytics* by George Mount (O'Reilly). Copyright 2021 George Mount, 978-1-492-09434-0."

If you feel your use of code examples falls outside fair use or the permission given above, feel free to contact us at *permissions@oreilly.com*.

O'Reilly Online Learning

 For more than 40 years, *O'Reilly Media* has provided technology and business training, knowledge, and insight to help companies succeed.

Our unique network of experts and innovators share their knowledge and expertise through books, articles, and our online learning platform. O'Reilly's online learning platform gives you on-demand access to live training courses, in-depth learning paths, interactive coding environments, and a vast collection of text and video from O'Reilly and 200+ other publishers. For more information, visit *http://oreilly.com*.

How to Contact Us

Please address comments and questions concerning this book to the publisher:

O'Reilly Media, Inc.
1005 Gravenstein Highway North
Sebastopol, CA 95472
800-998-9938 (in the United States or Canada)
707-829-0515 (international or local)
707-829-0104 (fax)

We have a web page for this book, where we list errata, examples, and any additional information. You can access this page at *https://oreil.ly/advancing-into-analytics*.

Email *bookquestions@oreilly.com* to comment or ask technical questions about this book.

For news and information about our books and courses, visit *http://oreilly.com*.

Find us on Facebook: *http://facebook.com/oreilly*.

Follow us on Twitter: *http://twitter.com/oreillymedia*.

Watch us on YouTube: *http://www.youtube.com/oreillymedia*.

Acknowledgments

First, I want to thank God for giving me this opportunity to cultivate and share my talents. At O'Reilly, Michelle Smith and Jon Hassell have been so enjoyable to work with, and I will be forever grateful for their offer to have me write a book. Corbin Collins kept me rolling during the book's development. Danny Elfanbaum and the production team turned the raw manuscript into an actual book. Aiden Johnson, Felix Zumstein, and Jordan Goldmeier provided invaluable technical reviews.

Getting people to review a book isn't easy, so I have to thank John Dennis, Tobias Zwingmann, Joe Balog, Barry Lilly, Nicole LaGuerre, and Alex Bodle for their comments. I also want to thank the communities who have made this technology and knowledge available, often without direct compensation. I've made some fantastic friends through my analytics pursuits, who have been so giving of their time and wisdom. My educators at Padua Franciscan High School and Hillsdale College made me fall in love with learning and with writing. I doubt I'd have written a book without their influence.

I also thank my mother and father for providing me the love and support that I'm so privileged to have. Finally, to my late Papou: thank you for sharing with me the value of hard work and decency.

## PART I

Foundations of Analytics in Excel

## CHAPTER 1

Foundations of Exploratory Data Analysis

"You never know what is gonna come through that door," Rick Harrison says in the opening of the hit show *Pawn Stars*. It's the same in analytics: confronted with a new dataset, you never know what you are going to find. This chapter is about *exploring* and *describing* a dataset so that we know what questions to ask of it. The process is referred to as *exploratory data analysis*, or EDA.

What Is Exploratory Data Analysis?

American mathematician John Tukey promoted the use of EDA in his book, *Exploratory Data Analysis* (Pearson). Tukey emphasized that analysts need first to *explore* the data for potential research questions before jumping into *confirming* the answers with hypothesis testing and inferential statistics.

EDA is often likened to "interviewing" the data; it's a time for the analyst to get to know it and learn about what interesting things it has to say. As part of our interview, we'll want to do the following:

- Classify our variables as continuous, categorical, and so forth
- Summarize our variables using descriptive statistics
- Visualize our variables using charts

EDA gives us a lot to do. Let's walk through the process using Excel and a real-life dataset. You can find the data in the *star.xlsx* workbook, which can be found in the *datasets* folder of this book's repository (*https://oreil.ly/VHslH*), under the *star* sub-folder. This dataset was collected for a study to examine the impact of class size on test scores. For this and other Excel-based demos, I suggest you complete the following steps with the raw data:

1. Make a copy of the file so that the original dataset is unchanged. We'll later be importing some of these Excel files into R or Python, so any changes to the datasets will impact that process.

2. Add an index column called *id*. This will number each row of the dataset so that the first row has an ID of 1, the second of 2, and so forth. This can be done quickly in Excel by entering numbers into the first few rows of the column, then highlighting that range and using Flash Fill to complete the selection based on that pattern. Look for the small square on the bottom right-hand side of your active cell, hover over it until you see a small plus sign, then fill out the rest of your range. Adding this index column will make it easier to analyze data by group.

3. Finally, convert your resulting dataset into a table by selecting any cell in the range, then going to the ribbon and clicking on Insert → Table. The keyboard shortcut is Ctrl + T for Windows, Cmd + T for Mac. If your table has headers, make sure the "My table has headers" selection is turned on. Tables carry quite a few benefits, not the least of which is their aesthetic appeal. It's also possible to refer to columns by name in table operations.

You can give the table a specific name by clicking anywhere inside it, then going to the ribbon and clicking Table Design → Table Name under the Properties group, as shown in Figure 1-1.

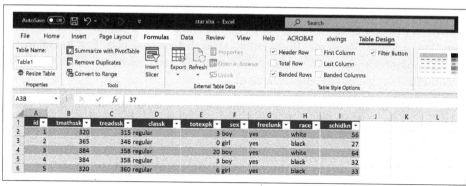

Figure 1-1. The Table Name box

Doing these first few analysis tasks will be good practice for other datasets you want to work with in Excel. For the *star* dataset, your completed table should look like Figure 1-2. I've named my table star. This dataset is arranged in a rectangular shape of columns and rows.

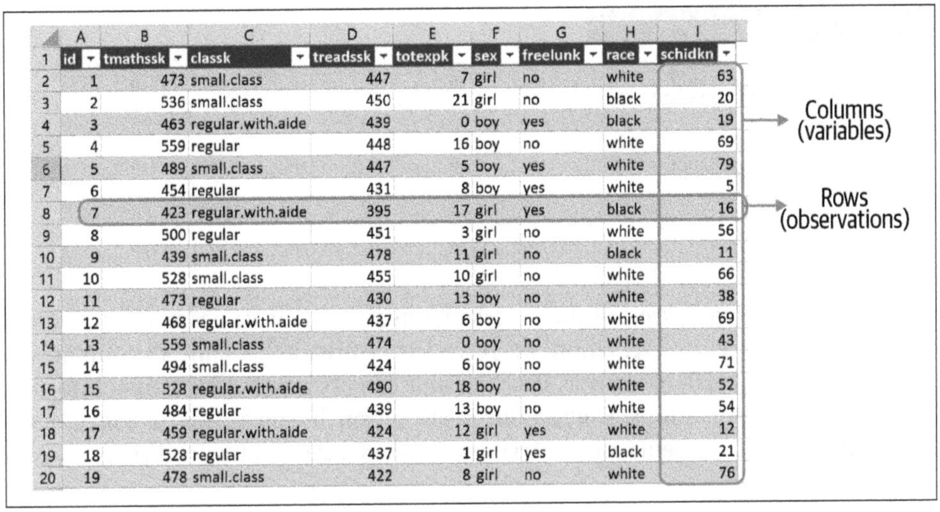

Figure 1-2. The star dataset, arranged in rows and columns

You've probably worked with enough data to know that this is a desirable shape for analysis. Sometimes, we need to clean up our data to get it to the state we want; I will discuss some of these data-cleaning operations later in the book. But for now, let's count our blessings and learn about our data and about EDA.

In analytics, we often refer to *observations* and *variables* rather than *rows* and *columns*. Let's explore the significance of these terms.

Observations

In this dataset we have 5,748 rows: each is a unique observation. In this case, measurements are taken at the student level; observations could be anything from individual citizens to entire nations.

Variables

Each column offers a distinct piece of information about our observations. For example, in the *star* dataset we can find each student's reading score (*treadssk*) and which class type the student was in (*classk*). We'll refer to these columns as *variables*. Table 1-1 describes what each column in *star* is measuring:

Table 1-1. Descriptions of the star dataset's variables

Column	Description
id	Unique identifier/index column
tmathssk	Total math scaled score
treadssk	Total reading scaled score

Column	Description
classk	Type of class
totexpk	Teacher's total years of experience
sex	Sex
freelunk	Qualified for free lunch?
race	Race
schidkn	School indicator

Ready for a tautology? We call them variables because their values may *vary* across observations. If every observation we recorded returned the same measurements, there wouldn't be much to analyze. Each variable can provide quite different information about our observations. Even in this relatively small dataset, we have text, numbers, and yes/no statements all as variables. Some datasets can have dozens or even hundreds of variables.

It can help to classify these variable types, as these distinctions will be important when we continue our analysis. Keep in mind that these distinctions are somewhat arbitrary and may change based on the purpose and circumstances of our analysis. You will see that EDA, and analytics in general, is highly iterative.

 Classifying variables is somewhat arbitrary and, like much of analytics, built on rules of thumb rather than hard-and-fast-criteria.

I will discuss the different variable types as shown in Figure 1-3, then classify the *star* dataset based on these distinctions.

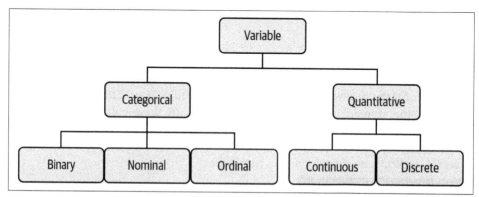

Figure 1-3. Types of variables

There are further types of variables that could be covered here: for example, we won't consider the difference between interval and ratio data. For a closer look at variable types, check out Sarah Boslaugh's *Statistics in a Nutshell*, 2nd edition (O'Reilly). Let's work our way down Figure 1-3, moving from left to right.

Categorical variables

Sometimes referred to as *qualitative* variables, these describe a quality or characteristic of an observation. A typical question answered by categorical variables is "Which kind?" Categorical variables are often represented by nonnumeric values, although this is not always the case.

An example of a categorical variable is country of origin. Like any variable, it could take on different values (United States, Finland, and so forth), but we aren't able to make quantitative comparisons between them (what is two times Indonesia, anyone?). Any unique value that a categorical variable takes is known as a *level* of that variable. Three levels of a country of origin could be US, Finland, or Indonesia, for example.

Because categorical variables describe a quality of an observation rather than a quantity, many quantitative operations on this data aren't applicable. For example, we can't calculate the *average* country of origin, but we could calculate the *most common*, or the overall frequency count of each level.

We can further distinguish categorical values based on how many levels they can take, and whether the rank-ordering of those levels is meaningful.

Binary variables can only take two levels. Often, these variables are stated as yes/no responses, although this is not always the case. Some examples of binary variables:

- Married? (yes or no)
- Made purchase? (yes or no)
- Wine type? (red or white)

In the case of wine type, we are implicitly assuming that our data of interest only consists of red or white wine...but what happens if we also want to analyze rosé? In that case, we can no longer include all three levels and analyze the data as binary.

Any qualitiative variable with more than two levels is a nominal variable. Some examples include:

- Country of origin (US, Finland, Indonesia, and so forth)
- Favorite color (orange, blue, burnt sienna, and so forth)
- Wine type (red, white, rosé)

Note that something like an ID number is a categorical variable stated numerically: while we *could* take an average ID number, this figure is meaningless. Importantly, there is no *intrinsic ordering* of nominal variables. For example, *red* as a color can't inherently be ordered higher or lower than *blue*. Since intrinsic ordering isn't necessarily clear, let's look at some examples of its use.

Ordinal variables take more than two levels, where there *is* an intrinsic ordering between these levels. Some examples of ordinal variables:

- Beverage size (small, medium, large)
- Class (freshman, sophomore, junior, senior)
- Weekdays (Monday, Tuesday, Wednesday, Thursday, Friday)

Here, we *can* inherently order levels: senior is higher than freshman, whereas we can't say the same about red versus blue. While we can *rank* these levels, we can't necessarily quantify the *distance* between them. For example, the difference in size between a small and medium beverage may not be the same as that between a medium and large beverage.

Quantitative variables

These variables describe a measurable quantity of an observation. A typical question answered by quantitative variables is "How much?" or "How many?" Quantitative variables are nearly always represented by numbers. We can further distinguish between quantitative variables based on the number of values they can take.

Observations of a *continuous* variable can in theory take an infinite number of values between any two other values. This sounds complicated, but continuous variables are quite common in the natural world. Some examples:

- Height (within a range of 59 and 75 inches, an observation could be 59.1, 74.99, or any other value in between)
- pH level
- Surface area

Because we can make quantitative comparisons across observations of continuous variables, a fuller range of analyses apply to them. For example, it makes sense to take the average of continuous variables, whereas with categorical ones, it doesn't. Later in this chapter, you'll learn how to analyze continuous variables by finding their descriptive statistics in Excel.

On the other hand, observations of a *discrete* variable can take only a fixed number of countable values between any two values. Discrete variables are quite common in the social sciences and business. Some examples include:

- Number of individuals in a household (within a range of 1 and 10, an observation could be 2 or 5, but not 4.3)
- Units sold
- Number of trees in a forest

Often, when we are dealing with discrete variables with many levels, or many observations, we treat them as continuous for the fuller range of statistical analysis that affords. For example, you may have heard that the average US household has 1.93 children. We know that no family *actually* has such a number of children. After all, this is a *discrete* variable that comes in whole numbers. However, across many observations this claim can be a helpful representation of how many children are to be expected in a typical household.

But wait, there's more! In more advanced analytics, we will also often recalculate and blend variables: for example, we may take a *logarithmic transformation* of one variable so that it meets the assumptions of a given analysis, or we may extract the meaning of many variables into fewer using a method called *dimensionality reduction*. These techniques are beyond the scope of this book.

Demonstration: Classifying Variables

Using what you've learned so far, classify the *star* variables using the types covered in Figure 1-3. As you think through it, don't hesitate to investigate the data. I'll give you an easy way to do so here, and we'll walk through a more thorough process later in this chapter.

One quick way to get a sense of what type variables may be is by finding the number of unique values they take. This can be done in Excel by checking the filter preview. I've clicked on the drop-down arrow next to the *sex* variable in Figure 1-4 and found it takes only two distinct values. What kind of variable do you think this might be? Take a moment to walk through the variables using this or other methods.

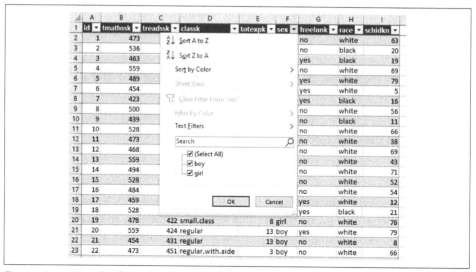

Figure 1-4. Using the filter preview to find how many distinct values a variable takes

Table 1-2 shows how I decided to classify these variables.

Table 1-2. How I classified these variables

Variable	Description	Categorical or quantitative?	Type?
id	Index column	Categorical	Nominal
tmathssk	Total math scaled score	Quantitative	Continuous
treadssk	Total reading scaled score	Quantitative	Continuous
classk	Type of class	Categorical	Nominal
totexpk	Teacher's total years of experience	Quantitative	Discrete
sex	Sex	Categorical	Binary
freelunk	Qualified for free lunch?	Categorical	Binary
race	Race	Categorical	Nominal
schidkn	School indicator	Categorical	Nominal

Some of these variables, like *classk* and *freelunk* were easier to categorize. Others, like *schidkn* and *id*, were not so obvious: they are stated in numeric terms, but cannot be quantitatively compared.

 Just because data is stated numerically doesn't mean it can be used as a quantitative variable.

You'll see that only three of these are quantitative: *tmathssk*, *treadssk*, and *totexpk*. I decided to classify the first two as continuous, and the last as discrete. To understand why, let's start with *totexpk*, the number of years of the teacher's experience. All of these observations are expressed in whole numbers, ranging from 0 to 27. Because this variable can only take on a fixed number of countable values, I classified it as discrete.

But what about *tmathssk* and *treadssk*, the test scores? These are also expressed in whole numbers: that is, a student can't receive a reading score of 528.5, only 528 or 529. In this respect, they are discrete. However, because these scores can take on so many unique values, in practice it makes sense to classify them as continuous.

It may surprise you to see that for such a rigorous field as analytics, there are very few hard-and-fast rules.

Recap: Variable Types

> Know the rules well, so you can break them effectively.
>
> —Dalai Lama XIV

The way we classify a variable influences how we treat it in our analysis—for example, we can calculate the mean of continous variables, but not nominal variables. At the same time, we often bend the rules for expediency—for example, taking the average of a discrete variable, so that a family has 1.93 children on average.

As we progress in our analysis, we may decide to twist more rules, reclassify variables, or build new variables entirely. Remember, EDA is an iterative process.

 Working with data and variables is an iterative process. The way we classify variables may change depending on what we find later in our exploration and the kinds of questions we decide to ask of our data.

Exploring Variables in Excel

Let's continue exploring the *star* dataset with *descriptive statistics* and *visualizations*. We will be conducting this analysis in Excel, although you could follow these same steps in R or Python and get matching results. By the end of the book, you'll be able to conduct EDA using all three methods.

We'll start our variable exploration with the categorical variables of *star*.

Exploring Categorical Variables

Remember that we are measuring *qualities* and not *quantities* with categorical variables, so these won't have a meaningful average, minimum, or maximum, for example. We can still conduct some analysis on this data, namely by counting *frequencies*. We can do this in Excel with PivotTables. Place your cursor anywhere in the *star* dataset and select Insert → PivotTable, as in Figure 1-5. Click OK.

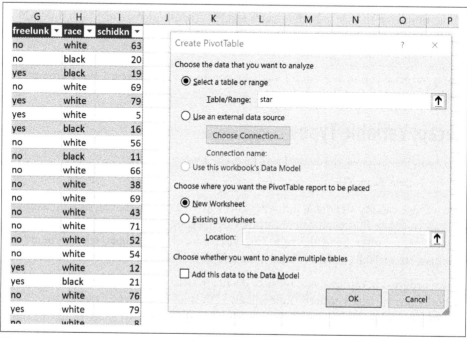

Figure 1-5. Inserting a PivotTable

I would like to find how many observations come from each class type. To do so, I will drag *classk* to the Rows area of the PivotTable, and *id* to the Values. By default, Excel will take the *sum* of the *id* field. It's made the mistake of assuming a categorical variable to be quantitative. We cannot quantitatively compare ID numbers, but we *can* count their frequencies. To do this on Windows, click "Sum of id" in the Values area and select Value Field Settings. Under "Summarize value field by," select "Count." Click OK. For Mac, click the *i* icon next "Sum of id" to do this. Now we have what we want: the number of observations for each class type. This is known as a *one-way frequency table* and is shown in Figure 1-6.

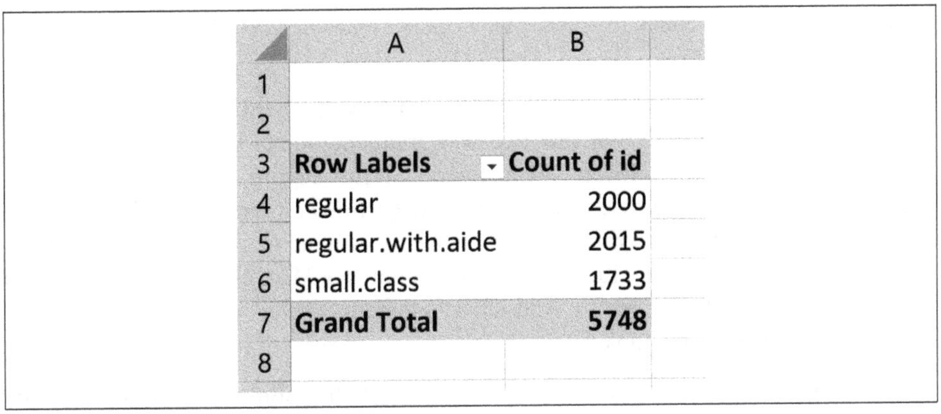

Figure 1-6. One-way frequency table of class type

Let's break this frequency count into observations of students who were and were not on the free lunch program. To do this, place *freelunk* into the Columns area of the PivotTable. We now have a *two-way* frequency table, as in Figure 1-7.

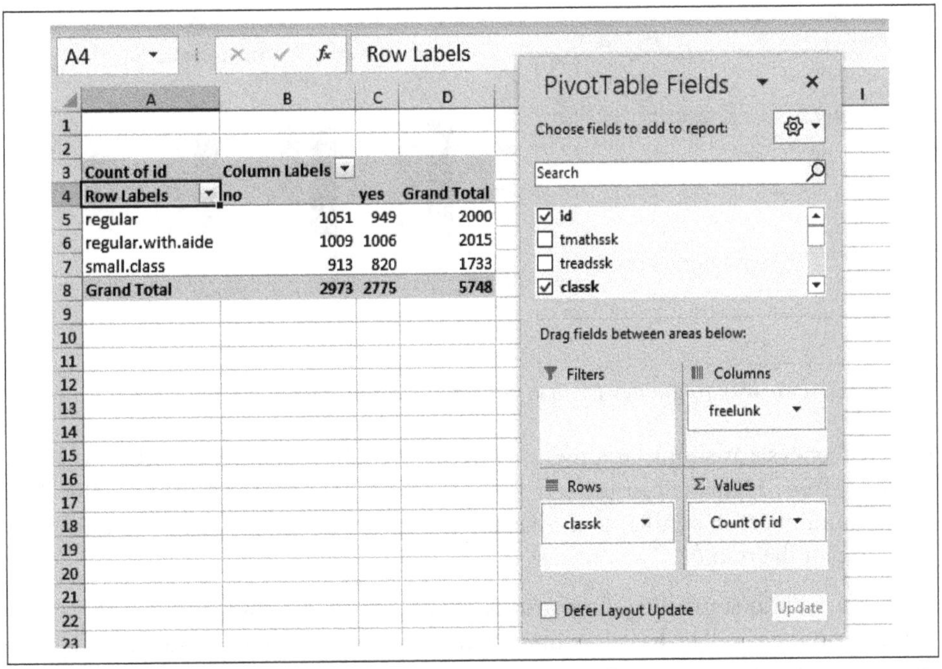

Figure 1-7. Two-way frequency table of class type by lunch program

Throughout this book, we'll be creating visualizations as part of our analysis. With everything else we have to cover, we won't spend too much time on the principles and techniques of data visualization. However, this field is well worth your study; for a helpful introduction, check out Claus O. Wilke's *Fundamentals of Data Visualization* (O'Reilly).

We can visualize a one- or two-way frequency table with a bar chart (also known as a *barplot* or *countplot*). Let's plot our two-way frequency table by clicking inside the PivotTable and clicking on Insert → Clustered Column. Figure 1-8 shows the result. I will add a title to the chart by clicking around its perimeter, then on the plus sign icon that appears on the upper right. Under the Chart Elements menu that appears, check on the selection for Chart Title. To find this menu on Mac, click on the chart and from the ribbon go to Design → Add Chart Element. I'll be adding charts this way several more times in the book.

Notice that the countplot and table have each split the number of observations by class type into students who are and are not on the free lunch program. For example, 1,051 and 949 indicate the first and second labels and bars on the table and countplot, respectively.

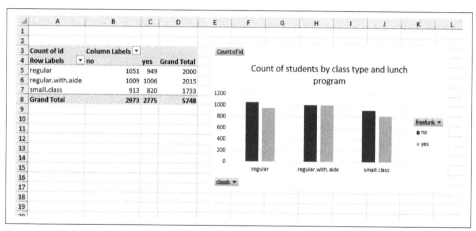

Figure 1-8. Two-way frequency table visualized as countplot

Even for analysis as simple as a two-way frequency table, it's not a bad idea to visualize the results. Humans can process lines and bars on a chart far more easily than they can numbers in a table, so as our analysis grows in complexity, we should continue to plot the results.

We can't make quantitative comparisons about categorical data, so any analysis we perform on them will be based on their counts. This may seem unexciting, but it's still important: it tells us what levels of values are most common, and we may want to compare these levels by other variables for further analysis. But for now, let's explore quantitative variables.

Exploring Quantitative Variables

Here, we'll run a fuller range of *summary* or *descriptive* statistics. Descriptive statistics allow you to summarize datasets using quantitative methods. Frequencies are one type of descriptive statistic; let's walk through some others and how to calculate them in Excel.

Measures of central tendency are one set of descriptive statistics that express what value or values a typical observation takes. We will cover the three most common of these measures.

First, the mean or average. More specifically the *arithmetic* mean, this is calculated by adding all observations together and dividing that number by the total number of observations. Of all the statistical measures covered, you may be most familiar with this one, and it's one we'll continue to refer to.

Next, the median. This is the observation found in the *middle* of our dataset. To calculate the median, sort or rank the data from low to high, then count into the data from both sides to find the middle. If two values are found in the middle, take the average to find the median.

Finally, the mode: the most commonly occurring value. It is also helpful to sort the data to find the mode. A variable can have one, many, or no modes.

Excel has a rich suite of statistical functions, including some to calculate measures of central tendency, which are shown in Table 1-3.

Table 1-3. Excel's functions for measuring central tendency

Statistic	Excel function
Mean	AVERAGE(number1, [number2], ...)
Median	MEDIAN(number1, [number2], ...)
Mode	MODE.MULT(number1, [number2], ...)

MODE.MULT() is a new function in Excel that uses the power of dynamic arrays to return multiple potential modes. If you do not have access to this function, try MODE(). Using these functions, find the measures of central tendency for our *tmathssk* scores. Figure 1-9 shows the results.

From this analysis, we see our three measures of central tendency have quite similar values, with a mean of 485.6, median of 484, and mode of 489. I've also decided to find out how often the mode occurs: 277 times.

	J	K	L	M
1		tmathssk central tendency		
2		Mean	485.6480515	=AVERAGE(star[tmathssk])
3		Median	484	=MEDIAN(star[tmathssk])
4		Mode	489	=MODE.MULT(star[tmathssk])
5		Mode -- how many?	277	=COUNTIF(star[tmathssk],L4)
6				

Figure 1-9. Calculating measures of central tendency in Excel

With all of these measures of central tendency, which one is right to focus on? I'll answer this with a brief case study. Imagine you are consulting at a nonprofit. You've been asked to look at donations and advise which measure of central tendency to track. The donations are shown in Table 1-4. Take a moment to calculate and decide.

Table 1-4. Consider which measure you should track given this data

$10 $10 $25 $40 $120

The mean seems like a conventional one to track, but is $41 *really* representative of our data? All individual donations but one were actually *less* than that; the $120 donation is inflating this number. This is one downside of the mean: extreme values can unduly influence it.

We wouldn't have this problem if we used the median: $25 is perhaps a better representation of the "middle value" than $41. The problem with this measure is it does not account for the precise value of each observation: we are simply "counting down" into the middle of the variable, without taking stock of each observation's relative magnitude.

That leaves us with the mode, which does offer useful information: the most *common* gift is $10. However, $10 is not all that representative of the donations as a whole. Moreover, as mentioned, a dataset can have multiple modes or none, so this is not a very stable measure.

Our answer to the nonprofit, then? It should track and evaluate them all. Each measure summarizes our data from a different perspective. As you will see in later chapters, however, it's most common to focus on the mean when conducting more advanced statistical analysis.

We will frequently analyze several statistics to get a fuller perspective about the same dataset. No one measure is necessarily better than others.

Now that we have established where the "center" of the variable is, we want to explore how "spread" those values are from the center. Several *measures of variability* exist; we'll focus on the most common.

First, the range, or the difference between the maximum and minimum values. Though simple to derive, it is highly sensitive to observations: just one extreme value, and the range can be misleading about where most observations are actually found.

Next, the variance. This is a measure of how spread observations are from the mean. This is a bit more intensive to calculate than what we've covered thus far. Our steps will be:

1. Find the mean of our dataset.
2. Subtract the mean from each observation. This is the *deviation*.
3. Take the sum of the squares of all deviations.
4. Divide the sum of the squares by the number of observations.

That's a lot to follow. For operations this involved, it can be helpful to use mathematical notation. I know it can take some getting used to and is intimidating at first, but consider the alternative of the previous list. Is that any more intelligible? Mathematical notation can provide a more precise way to express what to do. For example, we can cover all the steps needed to find the variance in Equation 1-1:

Equation 1-1. Formula for finding variance

$$s^2 = \frac{\Sigma (X - \bar{X})^2}{N}$$

s^2 is our variance. $(X - \bar{X})^2$ tells us that we need to subtract each observation X from the average $\bar{X}$, and square it. Σ tells us to sum those results. Finally, that result is divided by the number of observations N.

I will use mathematical notation a few more times in this book, but only to the extent that it is a more efficient way to express and understand a given concept than writing down all the steps discursively. Try calculating the variance of the numbers in Table 1-5.

Table 1-5. Measure the variability of this data

3 5 2 6 3 2

Because this statistic is comparatively more complex to derive, I will use Excel to manage the calculations. You'll learn how to calculate the variance using Excel's built-in functions in a moment. Figure 1-10 shows the results.

▲	A	B	C	D
1	observation	average	deviation	deviation squared
2	3	3.5	-0.5	0.25
3	5	3.5	1.5	2.25
4	2	3.5	-1.5	2.25
5	6	3.5	2.5	6.25
6	3	3.5	-0.5	0.25
7	2	3.5	-1.5	2.25
8				
9	sum of deviations squared	13.5 =SUM(D2:D7)		
10	number of observations	6 =COUNT(A2:A7)		
11	variance	2.25 =B9/B10		

Figure 1-10. Calculating the variance in Excel

You can find these results in the *variability* worksheet of the accompanying workbook for this chapter, *ch-1.xlsx*.

You may be asking why we are working with the *square* of the deviations. To see why, take the sum of the unsquared deviations. It's zero: these deviations cancel each other out.

The problem with the variance is that now we are working in terms of *squared deviations* of the original unit. This is not an intuitive way to analyze data. To correct for that, we'll take the square root of the variance, known as the *standard deviation*. Variability is now expressed in terms of the original unit of measurement, the mean. Equation 1-2 shows the standard deviation expressed in mathematical notation.

Equation 1-2. Formula for finding standard deviation

$$s = \sqrt{\frac{\Sigma\left(X_i - \bar{X}\right)^2}{N}}$$

Using this formula, the standard deviation of Figure 1-10 is 1.5 (the square root of 2.25). We can calculate these measures of variability in Excel using the functions in Table 1-6. Note that different functions are used for the *sample* versus *population* variance and standard deviation. The sample measure uses $N - 1$ rather than N in the denominator, resulting in a larger variance and standard deviation.

Table 1-6. Excel's functions for measuring variability

Statistic	Excel function
Range	MAX(number1, [number2], ...)_ - _MIN(number1, [number2], ...)
Variance (sample)	VAR.S(number1, [number2], ...)

Statistic	Excel function
Standard deviation (sample)	`STDEV.S(number1, [number2], ...)`
Variance (population)	`VAR.P(number1, [number2], ...)`
Standard deviation (population)	`STDEV.P(number1, [number2], ...)`

The distinction between the sample and population will be a key theme of later chapters. For now, if you're not sure you have collected *all* the data that you're interested in, use the *sample* functions. As you're beginning to see, we have *several* descriptive statistics to look out for. We can expedite calculating them using Excel's functions, but we can also use its Data Analysis ToolPak to derive a full suite of descriptive statistics with a few clicks.

 Some statistical measures differ when calculated for a population or a sample. If you're not sure which you're working with, assume the sample.

This add-in comes installed with Excel, but you need to load it first. For Windows, from the ribbon select File → Options > Add-ins. Then click Go toward the bottom of the menu. Select Analysis ToolPak from the menu, then click OK. It is not necessary to select the Analysis ToolPak–VBA option. For Mac, from the menu bar you will select Data → Analysis Tools. Select Analysis ToolPak from the menu, then click OK. You may need to restart Excel to complete the configuration. After that, you will see a new Data Analysis button in the Data tab.

In Table 1-1, we determined that *tmathssk* and *treadssk* are continuous variables. Let's calculate their descriptive statistics using the ToolPak. From the ribbon, select Data → Data Analysis → Descriptive Statistics. A menu will appear; select the input range `B1:C5749`. Make sure to turn the checkboxes on for "Labels in First Row" and "Summary statistics." Your menu should look like Figure 1-11. You can leave the other settings as-is and click OK.

This will insert descriptive statistics for these two variables into a new worksheet, as in Figure 1-12.

Now let's look at finding descriptive statistics for each level of a categorical variable for the sake of comparison across groups. To do this, insert a new PivotTable based off the *star* data into a new worksheet. Place *freelunk* in the Columns area, *id* in the Rows, and *Sum of treadssk* in the Values section. Remember that the *id* field is a unique identifier, so we really shouldn't sum this in the PivotTable, despite what it thinks.

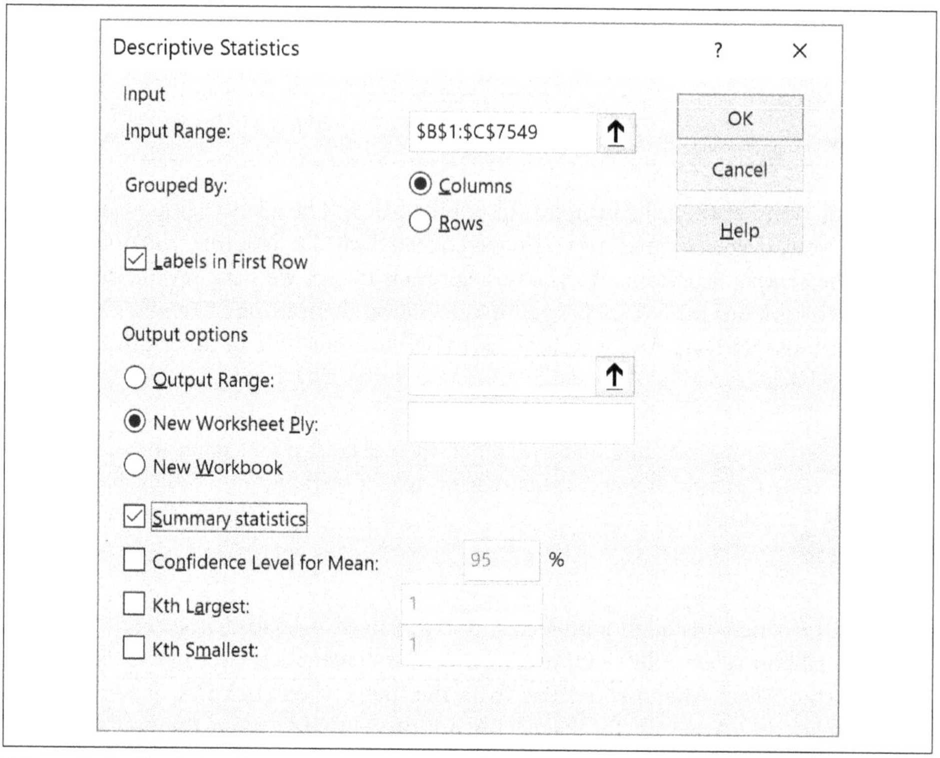

Figure 1-11. Deriving descriptive statistics with the Analysis ToolPak

	A	B	C	D
1	tmathssk		treadssk	
2				
3	Mean	485.6480515	Mean	436.7423452
4	Standard Error	0.63010189	Standard Error	0.419080917
5	Median	484	Median	433
6	Mode	489	Mode	437
7	Standard Deviation	47.77153121	Standard Deviation	31.77285677
8	Sample Variance	2282.119194	Sample Variance	1009.514427
9	Kurtosis	0.289321748	Kurtosis	3.83779705
10	Skewness	0.473937363	Skewness	1.340898831
11	Range	306	Range	312
12	Minimum	320	Minimum	315
13	Maximum	626	Maximum	627
14	Sum	2791505	Sum	2510395
15	Count	5748	Count	5748
16				

Figure 1-12. Descriptive statistics derived from the Analysis ToolPak

For this and any future PivotTable operations we'll be conducting, it's best to turn off all totals by clicking inside it and selecting Design → Grand Totals → Off for Rows and Columns. This way we don't mistakenly include grand totals as part of the analysis. You can now use the ToolPak to insert descriptive statistics. Figure 1-13 shows the result.

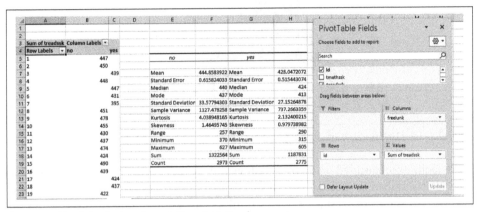

Figure 1-13. Calculating descriptive statistics by group

You know the majority of these measures already; this book will touch on the rest later. It may seem like all of the information presented by the ToolPak negates any need for visualizing the data. In fact, visualizations still play an indispensable role in EDA. In particular, we'll use them to tell us about the *distribution* of observations across the entire range of values in a variable.

First, we'll look at histograms. With these plots, we can visualize the relative frequency of observations by interval. To build a histogram of *treadssk* in Excel, select that range of data, then go to the ribbon and select Insert → Histogram. Figure 1-14 shows the result.

We can see from Figure 1-14 that the most frequently occurring interval is between 426.6 and 432.8, and there are approximately 650 observations falling in this range. None of our actual test scores include decimals, but our x-axis may include them depending on how Excel establishes the intervals, or bins. We can change the number of bins by right-clicking on the x-axis of the chart and selecting Format Axis. A menu will appear at the right. (These features are not available for Mac.)

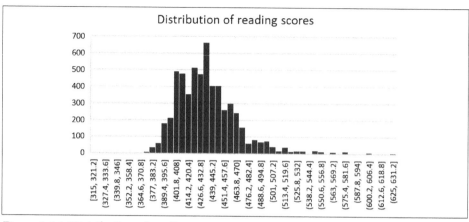

Figure 1-14. Distribution of reading scores

By default, Excel decided on 51 bins, but what if we (approximately) halved and doubled that number at 25 and 100, respectively? Adjust the numbers in the menu; Figure 1-15 shows the results. I like to think of this as "zooming in and out" on the details of the distribution.

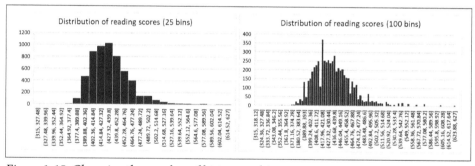

Figure 1-15. Changing the number of histogram bins

With the distribution visualized as a histogram, we can quickly see that there are a sizable number of test scores to the extreme right of the distribution, but that most test scores are overwhelmingly in the 400–500 range.

What if we wanted to see how the distribution of reading scores varies across the three class sizes? Here, we are comparing a continuous variable across three levels of a categorical one. Setting this up with a histogram in Excel will take some "hacking," but we can lean on PivotTables to get the job done.

Insert a new PivotTable based on the *star* dataset, then drag *treadssk* to the Rows area, *classk* to the Columns area, and "Count of id" to the Values area. Again, subsequent analysis will be easier if we remove totals from the PivotTable.

Now let's create a chart from this data. Click anywhere in your PivotTable, and from the ribbon, select Insert → Clustered Column. The result, shown in Figure 1-16, is extremely hard to read, but compare it to the source PivotTable: it is telling us that for students with a score of 380, 10 had regular classes, 2 had regular classes with aides, and 2 had small classes.

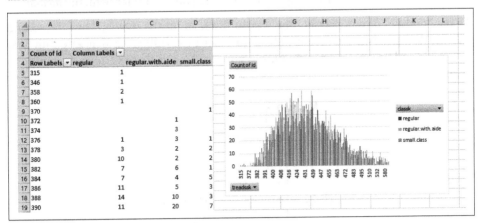

Figure 1-16. Starting a multigroup histogram

From here, it's a matter of rolling these values up into larger intervals. To do this, right-click anywhere inside the values of your PivotTable's first column and select Group. Excel will default this grouping to increments of 100; change it to 25.

A recognizable histogram is starting to emerge. Let's reformat the chart to make it look even more like one. Right-click on any of the bars of the chart and select Format Data Series. You will turn Series Overlap to 75% and Gap Width to 0%. Figure 1-17 shows the result.

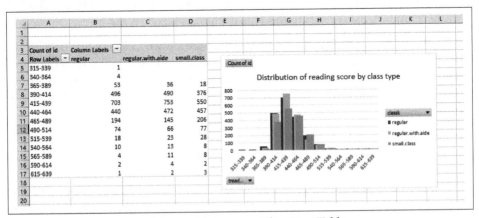

Figure 1-17. Creating a multigroup histogram with a PivotTable

We could set the gap widths to completely intersect, but then it becomes even harder to see the regular class size distribution. Histograms are a go-to visualization to see the distribution of a continuous variable, but they can quickly get cluttered.

As an alternative, let's look at boxplots. Here, we'll visualize our distribution in terms of *quartiles*. The center of the boxplot is a measure you're familiar with, the *median*.

As the "middle" of our dataset, one way to think about the median is as the second quartile. We can find the first and third quartiles by dividing our dataset evenly into quadrants and finding their midpoints. Figure 1-18 labels these various elements of a boxplot.

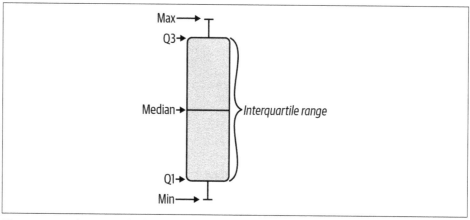

Figure 1-18. Elements of a boxplot

The part of the resulting plot found in the "box" is known as the *interquartile range*. This range is used as the basis for deriving other parts of the plot. The remaining range that falls within 1.5 times the interquartile range is represented by two lines or "whiskers." In fact, Excel refers to this type of chart as Box & Whisker.

Observations that aren't found within this range are shown as individual dots on the plot. These are considered *outliers*. The boxplot may be more complex than the histogram, but fortunately Excel will handle all preparation for us. Let's return to our *treadssk* example. Highlight this range, then from the ribbon select Insert → Box & Whisker.

We can see in Figure 1-19 that our interquartile range falls within about 415 to 450, and that there are several outliers, especially on the high side. We noticed similar patterns about the data from the histogram, although we had a more visual perspective of the complete distribution, and were able to examine at different levels of granularity with different bin widths. Just like with the descriptive statistics, each visualization offers a unique perspective of the data; none is inherently superior to the others.

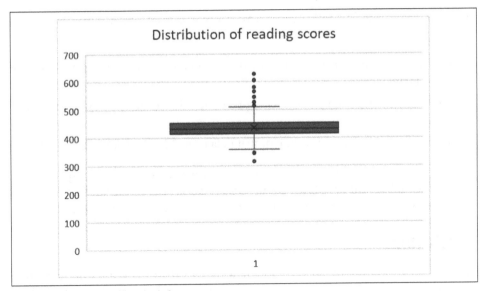

Figure 1-19. A boxplot of reading scores

One advantage of the boxplot is that it gives us some precise information about where the quartiles of our data are located, and what observations are considered outliers. Another is that it can be easier to compare distributions across multiple groups. To make boxplots of multiple groups in Excel, it's easiest to have the categorical variable of interest directly to the left of the continuous one. In this manner, move *classk* to the left of *treadssk* in your data source. With this data selected, click Insert → Box & Whisker from the ribbon. In Figure 1-20 we see that the general distribution of scores looks similar across the three groups.

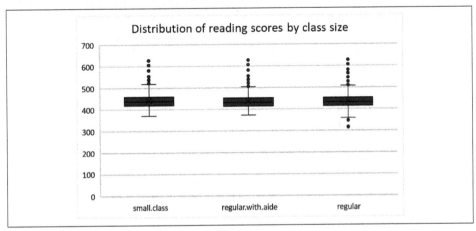

Figure 1-20. A boxplot of reading scores by class type

To recap, when working with quantitative data we can do much more than count frequencies:

- We can determine what value(s) the data is centered around using measures of central tendency.
- We can determine how relatively spread out that data is using measures of variability.
- We can visualize the distribution of that data using histograms and boxplots.

There are other descriptive statistics and other visualizations to explore quantitative variables with. You will even learn about some of them later in the book. But this is a good start with the most crucial questions to ask of your data during EDA.

Conclusion

While we never know what we'll get in a new dataset, the EDA framework gives us a good process to make sense of it. We now know what kind of variables we're working with in *star*, and how their observations as a whole look and behave: quite an in-depth interview. In Chapter 3, we will build on this work by learning how to *confirm* the insights we've gained about the data by exploring it. But before that, we'll take a tour of probability in Chapter 2, which provides much of the fuel for the analytics engine.

Exercises

Practice your EDA skills with the *housing* dataset, available in the book's repository (*https://oreil.ly/LHiLl*) under *datasets* → *housing* → *housing.xlsx*. This is a real-life dataset consisting of housing sales prices in the city of Windsor, Ontario, Canada. You can find a description of the variables on the *readme* worksheet of the file. Complete the following, and don't hesitate to complete your own EDA as well:

1. Classify each variable's type.
2. Build a two-way frequency table of *airco* and *prefarea*.
3. Return descriptive statistics for *price*.
4. Visualize the distribution of *lotsize*.

You can find a solution to these and all other book exercises in the *exercise-solutions* folder of the book repository. There is a file named for each chapter.

## CHAPTER 2

Foundations of Probability

Have you ever stopped to consider what your meteorologist really means by a 30% chance of rain? Barring a crystal ball, they can't say for sure that it will rain. That is, they are *uncertain* about an *outcome*. What they *can* do is *quantify* that uncertainty as a value between 0% (certain it will not rain) and 100% (certain it will rain).

Data analysts, like meteorologists, do not possess crystal balls. Often, we want to make claims about an entire population while only possessing the data for a sample. So we too will need to quantify uncertainty as a probability.

We'll start this chapter by digging deeper into how probability works and how probabilities are derived. We'll also use Excel to simulate some of the most important theorems in statistics, which are largely based on probability. This will put you on excellent footing for Chapter 3 and Chapter 4, where we'll perform inferential statistics in Excel.

Probability and Randomness

Colloquially, we say that something is "random" when it seems out of context or haphazard. In probability, something is *random* when we know an event will *have* an outcome, but we're not sure what that outcome will be.

Take a six-sided die, for example. When we toss the die, we know it will land on one side—it won't disappear or land on multiple sides. Knowing that we'll get *an* outcome, but not *which* outcome, is what's meant by randomness in statistics.

Probability and Sample Space

We know that when the die lands, it will display a number between one and six. This set of all outcomes is called a *sample space*. Each of these outcomes is assigned a probability greater than zero, because it's possible that the die may land on any of its sides. Summed together, these probabilities come to 1, because we are certain the outcome will be one of these possibilities in the sample space.

Probability and Experiments

We've determined that rolling a die is random, and we've outlined its *sample space*. We can now begin to build experiments for this random event. In probability, experiments are procedures that can be infinitely replicated with a consistent sample space of possible outcomes.

Some experiments take many years of planning, but ours is fortunately simple: roll a die. Each time we do, we get another value between one and six. The result is our outcome. Each die roll is known as a *trial* of the experiment.

Unconditional and Conditional Probability

Given what we know about probability so far, a typical probabilistic question about die rolls might be: "What is the probability of rolling a four?" This is called the *marginal* or *unconditional* probability, as we are only looking at one event in isolation.

But what about a question like "What is the probability of rolling a two, given that we rolled a one in the last trial?" To answer this, we would be discussing *joint* probability. Sometimes when we are studying the probability of two events, we know the outcome of one but not the other. This is known as *conditional* probability, and one way to calculate it is with Bayes' rule.

We will not cover Bayes' rule, and the many areas of probability and statistics that apply it, in this book, but it's well worth your future study. Check out Will Kurt's *Bayesian Statistics the Fun Way* (No Starch Press) for a fantastic introduction. You will see that Bayesianism offers a unique approach to working with data with some impressive applications for analytics.

 The schools of thought developed around Bayes' rule diverge from the so-called frequentist approaches used in this book and much of classical statistics.

Probability Distributions

So far, we've learned what makes our die toss a random experiment, and we've enumerated the sample space of what possible values a trial might take. We know that the sum of the probabilities of each outcome must equal 1, but what is the relative probability for each outcome? For this we can refer to *probability distributions*. A probability distribution is a listing of what possible outcomes an event can take, and how common each outcome is. While a probability distribution could be written as a formal mathematical function, we will instead focus on its quantitative output.

In Chapter 1, you learned about the difference between discrete and continuous variables. There are also related discrete and continuous probability *distributions*. Let's learn more, starting with the former.

Discrete Probability Distributions

We'll continue with our example of a die toss. This is considered a *discrete* probability distribution because there are a countable number of outcomes: for example, while a die toss can result in a 2 or a 3, it can never result in a 2.25.

In particular, a die toss is a *discrete uniform* probability distribution because each outcome is equally likely for any trial: that is, we are just as likely to roll a 4 as we are a 2, and so forth. To be more specific, we have a one-in-six probability for each outcome.

To follow along with this and the other Excel demos in this chapter, head to the *ch-2.xlsx* file in this book's repository (*https://oreil.ly/1hlYj*). For most of these exercises, I completed some staging of the worksheet already and will work through the remainder with you here. Start with the *uniform-distribution* worksheet. Each possible outcome X is listed in the range A2:A7. We know that it's equally likely to get any outcome, so our formula in B2:B7 should be =1/6. $P(X=x)$ indicates the probability that a given event will result in the listed outcome.

Now, select the range A1:B7 and from the ribbon, go to Insert > Clustered Column. Your probability distribution and visualization should look like Figure 2-1.

Welcome to your first, if unexciting, probability distribution. Notice the gaps between values in our visualization? This is a wise choice to indicate that these are *discrete* and not continuous outcomes.

Sometimes we may want to know the *cumulative* probability of an outcome. In this case, we take a running total of all probabilities until we reach 100% (because the sample space must sum to 1). We'll find the probability of an event being *less than or equal to* a given outcome in column C. We can set up a running total in the range C2:C7 with the formula =SUM($B$2:B2).

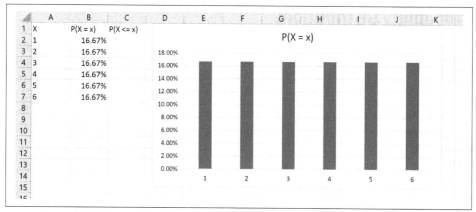

Figure 2-1. The probability distribution of a six-sided die toss

Now, select the range A1:A7, hold down the Ctrl key for Windows or Cmd for Mac, and highlight C1:C7. With this noncontiguous range selected, create a second clustered column chart. Do you see the difference between a probability distribution and a *cumulative* probability distribution in Figure 2-2?

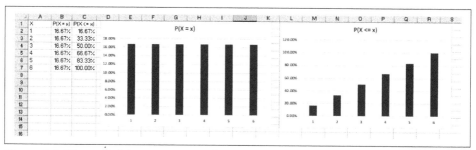

Figure 2-2. The probability versus cumulative probability distribution of a six-sided die toss

Based on logic and mathematical reasoning, we've been assuming a one-in-six probability of landing on any side of the die. This is called our *theoretical probability*. We could also find the probability distribution empirically by rolling a die many times and recording the results. This is called an *experimental probability*. After all, we could find through experiments that the probability for each side of the die is really *not* one in six as theoretically reasoned, and that the die is biased toward a given side.

We have a couple of options for deriving experimental probabilities: first, we could indeed conduct a real experiment. Of course, rolling a die dozens of times and recording the results might get quite tedious. Our alternative is to get the computer to do the heavy lifting, and *simulate* the experiment. Simulation often provides a decent approximation of reality, and is frequently used when running experiments in real life is too difficult or time-consuming. The downside of simulation is that it can fail to reflect any anomalies or idiosyncrasies from the real-life experiment that it intends to represent.

Simulation is frequently used in analytics to glean what might happen in real life when finding out through actual experiments is too difficult or even impossible.

To simulate the experiment of rolling a die, we need a way to consistently choose a number between one and six at random. We can do this using Excel's random number generator, RANDBETWEEN(). The results you see in the book will be different than what you get when you try it yourself...but they will *all* be random numbers between one and six.

Your individual results using Excel's random number generator will look different from what's been recorded in the book.

Now, go to the *experimental-probability* worksheet. In column A, we have labeled 100 die toss trials for which we'd like to record the outcome. At this point, you could start rolling a real die and recording the results in column B. Your more efficient, if less realistic, alternative is to simulate the results with RANDBETWEEN().

This function takes two arguments:

```
RANDBETWEEN(bottom, top)
```

We are using a six-sided die, which makes our range go between one and six:

```
RANDBETWEEN(1, 6)
```

RANDBETWEEN() only returns whole numbers, which is what we want in this case: again, this is a *discrete* distribution. Using the fill handle, you can generate an outcome for all 100 trials. Don't get too attached to your current outcomes: press F9 in Windows, fn-F9 for Mac, or from the ribbon select Formulas → Calculate Now. This will recalculate your workbook, and regenerate your random numbers.

Let's compare our theoretical versus experimental probabilities of a die toss in columns D-F. Column D will be used to enumerate our sample space: the numbers one through six. In column E, take the theoretical distribution: 1/6, or 16.67%. In column F, calculate the experimental distribution from columns A and B. This is the percentage of times we find each outcome across all trials. You can find this using the formula:

```
=COUNTIF($B$2:$B$101, D2)/COUNT($A$2:$A$101)
```

Select your range D1:F7 and from the ribbon go to Insert → Clustered Column. Your worksheet should now look like Figure 2-3. Try recalculating it a couple of times.

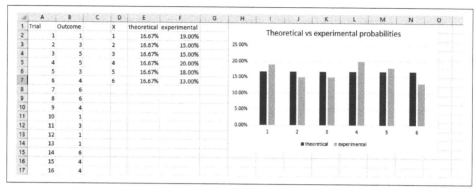

Figure 2-3. Theoretical versus experimental probabilities of a six-sided die toss

It looks like, based on our experimental distribution, we were right to predict an equal likelihood of rolling any number. Of course, our experimental distribution isn't *perfectly* like the theoretical: there will always be some error due to random chance.

It could be the case, however, that were we to conduct the experiment in real life, the results would differ from what we derived from simulation. Perhaps the real-life die of interest is *not* fair, and we've overlooked that by relying on our own reasoning and Excel's algorithm. It seems like a trivial point, but often probabilities in real life don't behave as we (or our computers) expect them to.

The discrete uniform is one of many discrete probability distributions; others commonly used in analytics include the binomial and Poisson distributions.

Continuous Probability Distributions

A distribution is considered continuous when an outcome can take any possible value between two other values. We will focus here on the normal distribution, or the *bell curve*, as depicted with a histogram. You may be familiar with this famous shape, seen in Figure 2-4.

You'll see in this chart a perfectly symmetrical distribution centered around the variable's mean (μ). Let's dig in on what the normal distribution is and what it tells us, using Excel to illustrate the fundamental statistical concepts that are based on it.

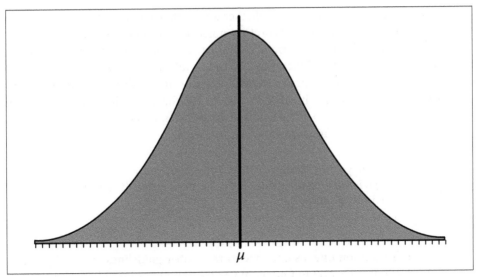

Figure 2-4. The normal distribution depicted with a histogram

The normal distribution is worth reviewing in part because it is so common in the natural world. For example, Figure 2-5 shows histograms of the distribution of student heights and wine pH levels, respectively. These datasets are available in the book's repository (*https://oreil.ly/1hlYj*) for you to explore in the *datasets* folder under *heights* and *wine*, respectively.

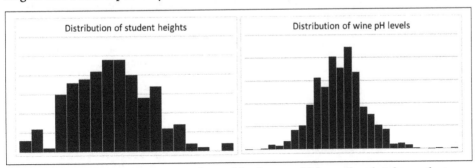

Figure 2-5. Two normally distributed variables from real life: student heights and wine pH levels

You may be wondering how we know when a variable is normally distributed. Good question. Think back to our die toss example: we enumerated all possible outcomes, derived a theoretical distribution, then derived an experimental distribution (via simulation) to compare the two. Consider the histograms in Figure 2-5 as *experimental distributions* of their own: in this case, the data was collected manually rather than relying on a simulation.

There are several ways to determine whether a real-life dataset with its experimental distribution is close enough to a theoretical normal distribution. For now, we'll look out for the telltale bell curve histogram: a symmetrical shape, with the majority of values found near the center. Other ways include evaluating skewness and kurtosis, which are two additional summary statistics measuring the distribution's symmetry and peakedness, respectively. It's also possible to test for normality using methods of statistical inference. You'll learn the basics of statistical inference in Chapter 3. But for now, we'll go by the rule: "You know it when you see it."

 When you are dealing with real-life data, you are dealing with *experimental* distributions. They will never perfectly match theoretical distributions.

The normal distribution offers some easy-to-remember guidelines for what percentage of observations we expect to find within a given number of standard deviations of the mean. Specifically, for a normally distributed variable we expect:

- 68% of observations to fall within one standard deviation of the mean
- 95% of observations to fall within two standard deviations of the mean
- 99.7% of observations to fall within three standard deviations of the mean

This is known as the *empirical rule*, or the *68–95–99.7 rule*. Let's see it in action using Excel. Next, go to the *empirical-rule* worksheet, as shown in Figure 2-6.

	A	B	C	D	E	F	G	H	I	J	K	L	M	N
1	Mean	50								Empirical rule: 1 s.d.				
2	Standard deviation	10												
3							100.00%							
4	% of observations	0.00000%	0.00000%	0.00000%	0.00000%		90.00%							
5							80.00%							
6			1	2	3		70.00%							
7		lower	40	30	20		60.00% / 50.00%							
8		upper	60	70	80		40.00%							
9	x	P(x)	1 s.d.	2 s.d.	3 s.d.		30.00%							
10	1						20.00%							
11	2						10.00%							
12	3						0.00%							
13	4													
14	5								■ P(x) ■ 1 s.d.					
15	6													
16	7													
17	8								Empirical rule: 2 s.d.					
18	9						100.00%							
19	10						90.00%							

Figure 2-6. The start of the empirical-rule worksheet

In cells A10:A109 we have the values 1–100. Our objective in cells B10:B109 is to find what percentage of observations will take on these values for a normally distributed variable with a mean of 50 and a standard deviation of 10 (cells B1 and B2, respectively). We will then find what percentage of observations fall within one, two, and three standard deviations of the mean in `C10:E109`. Once we do so, the charts to the right will be populated. Cells `C4:E4` will also find the total percentages for each column.

The normal distribution is continuous, which means that observations can theoretically take on any value between two other values. This makes for a *lot* of outcomes to assign a probability to. For simplicity, it's common to bin these observations into discrete ranges. The probability mass function (PMF) will return the probabilities found for each discrete bin in the range of observations. We'll use Excel's `NORM.DIST()` function to calculate a PMF for our variable in the range 1–100. This function is more involved than others used so far, so I've described each argument in Table 2-1.

Table 2-1. The arguments needed for `NORM.DIST()`

Argument	Description
X	The outcome for which you want to find the probability
Mean	The mean of the distribution
Standard_dev	The standard deviation of the distribution
Cumulative	If TRUE, a cumulative function is returned; if FALSE, the mass function is returned

Column A of our worksheet contains our outcomes, `B1` and `B2` contain our mean and standard deviation, respectively, and we want a mass instead of a cumulative distribution. Cumulative would return a running sum of the probability, which we don't want here. That makes our formula for `B10`:

```
=NORM.DIST(A10, $B$1, $B$2, 0)
```

Using the fill handle, you'll get the percentage likelihood of an observation taking on each value from 0 to 100. For example, you'll see in cell `B43` that there is approximately a 1.1% chance of an observation being equal to 34.

We can see in cell `B4` that an outcome is likely to be between 1 and 100 over 99.99% of the time. Importantly, this number is not equal to 100%, because an observation in a continuous distribution can take on *any* possible value—not just those from 1 to 100. In cells `C7:E8` I've written formulas to find the range of values within one, two and three standard deviations of our mean.

We can use these thresholds along with conditional logic to find what parts of our probability mass function in column B can be found within these respective regions. In cell `C10` enter the following formula:

```
=IF(AND($A10 > C$7, $A10 < C$8), $B10, "")
```

This function will carry the probability over from column B if the value of column A falls within the standard deviation range. If it falls outside the range, the cell is left blank. Using the fill handle, you can apply this formula to the entire range C10:E109. Your worksheet should now look like Figure 2-7.

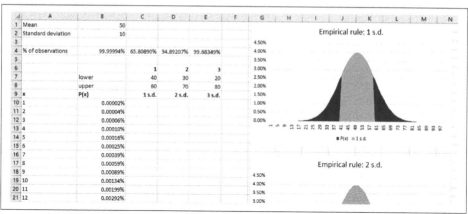

Figure 2-7. The empirical rule demonstrated in Excel

Cells C4:E4 indicate that we find approximately 65.8%, 94.9%, and 99.7% of values within one, two, and three standard deviations of the mean, respectively. These numbers are very close to matching the 68–95–99.7 rule.

Now, take a look at the resulting visualizations: we see that a sizable majority of observations can be found within one standard deviation, and more still within two. By three standard deviations, it's hard to see the part of the chart in Figure 2-8 that is *not* covered, but it's still there. (Remember, this is only 0.3% of all observations.)

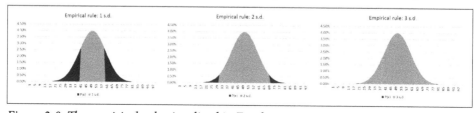

Figure 2-8. The empirical rule visualized in Excel

What happens when you change the standard deviation of our example to eight? To 12? The bell curve shape remains symmetrically centered around the mean of 50 but contracts and expands: a lower standard deviation leads to a "tighter" curve and vice versa. In any case, the empirical rule roughly applies to the data. If you shift the mean to 49 or 51, you see the "center" of the curve move along the x-axis. A variable can have any mean and standard deviation and still be normally distributed; its resulting probability mass function will be different.

Figure 2-9 shows two normal distributions with different means and standard deviations. Despite their very different shapes, they both still follow the empirical rule.

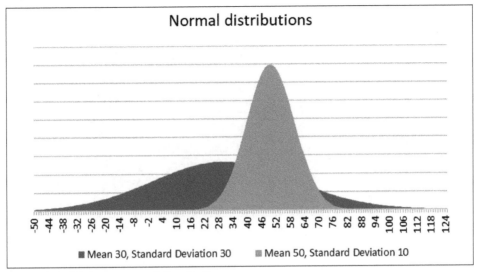

Figure 2-9. *Different normal distributions*

 A normal distribution can have any possible combination of mean and standard deviation. The resulting probability density function will change, but will roughly follow the empirical rule.

The normal distribution is also important because of its place in the central limit theorem. I call this theorem the "missing link of statistics" for reasons you'll see in this and following chapters.

For an example of the central limit theorem, we will use another common game of chance: roulette. A European roulette wheel is equally likely to return any number between 0 and 36 (compared to the American wheel, which has slots labeled both 0 and 00). Based on what you've learned about die tosses, what kind of probability distribution is this? It's a discrete uniform. Does it seem odd that we're analyzing this distribution in a demo I said was about the normal distribution? Well, we have the central limit theorem to thank here. To see this theorem in action for yourself, go to the *roulette-dist* worksheet and simulate 100 spins of a roulette wheel in B2:B101 using RANDBETWEEN():

```
RANDBETWEEN(0, 36)
```

Visualize the result using a histogram. Your worksheet should look like Figure 2-10. Try recalculating a few times. You will see that each time you get a rather flat-looking histogram. This is indeed a discrete uniform distribution, where it's equally likely to land on any number between 0 and 36.

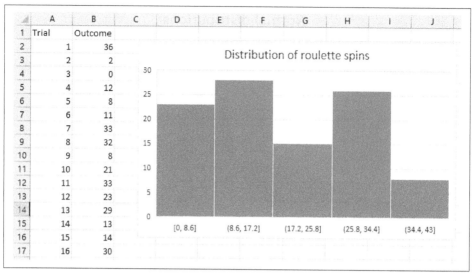

Figure 2-10. The distribution of simulated roulette spins

Now go to the *roulette-sample-mean-dist* worksheet. Here we will be doing something a little different: we'll simulate 100 spins, then take the average of those spins. We'll do this 100 times and plot the distribution of these trial averages, again as a histogram. This "average of averages" is known as a *sample mean*. Once you've done that using the RANDBETWEEN() and AVERAGE() functions, you should see something like Figure 2-11.

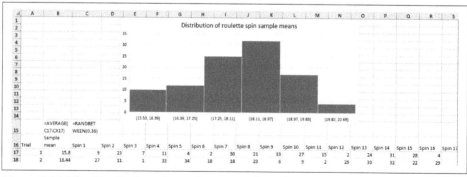

Figure 2-11. The distribution of sample means of simulated roulette spins

This distribution no longer looks like a rectangle: in fact, it looks like a bell curve. It's symmetrical, and most observations are clustered around the center: we now have a normal distribution. How could our distribution of sample means be normally distributed, when the roulette spins themselves are not? Welcome to the very special kind of magic known as the *central limit theorem* (CLT).

Put formally, the CLT tells us that:

> The distribution of sample means will be normal or nearly normal if the sample size is large enough.

This phenomenon is a game changer, because it allows us to use the unique traits of the normal distribution (such as the empirical rule) to make claims about the sample means of a variable, even when that variable itself isn't normally distributed.

Did you catch the fine print? The CLT only applies *when the sample size is large enough*. It's an important disclaimer, but also an ambiguous one: how large is large enough? Let's gather some ideas with another Excel demo. Head to the *law-of-large-numbers* worksheet to follow along. In column B, we can simulate the outcomes of 300 trials of a roulette spin using RANDBETWEEN(0, 36).

In column C, we want to take a running average of the outcome. We can do so using mixed references; in column C enter the following and drag it alongside your 300 trials:

```
=AVERAGE($B$2:B2)
```

This will result in finding the running average of column B. Select your resulting data in column C and then head to the ribbon and click Insert → Line. Take a look at your line chart and recalculate the workbook a few times. Each resulting simulation will be different than what's shown in Figure 2-12, but as a pattern the average tends to converge to 18 with more spins, which makes sense: it's the average between 0 and 36. This anticipated figure is known as the *expected value*.

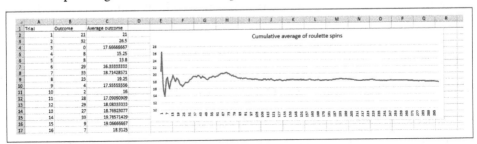

Figure 2-12. The law of large numbers, visualized in Excel

This phenomenon is known as the *law of large numbers* (LLN). Stated formally, the LLN tells us:

> The average of results obtained from trials become closer to the expected value as more trials are performed.

This definition, though, begs the question we first asked: how large does a sample size need to be for the CLT to apply? You'll often hear 30 given as a threshold. More conservative guidelines may call for a sample size of 60 or 100. Given these sample size guidelines, look back to Figure 2-12. See how it indeed settles more closely to the expected value at around these thresholds?

 The law of large numbers provides a loose rule of thumb for an adequate sample size that meets the CLT.

Sample sizes of 30, 60, and 100 are purely rules of thumb; there are more rigorous ways to determine what sample sizes are needed for the CLT to apply. For now, remember this: given that our sample size meets these thresholds, our sample mean should be close to the expected value (thanks to the LLN), and should also be normally distributed (thanks to the CLT).

Several continuous probability distributions exist, such as the exponential and triangular. We focused on the normal distribution both because of its ubiquity in the real world and because of its special statistical properties.

Conclusion

As mentioned at the beginning of this chapter, data analysts live in a world of uncertainty. Specifically, we often want to make claims about an entire population, while only possessing a sample's worth of data. Using the framework of probability covered in this chapter, we will be able to do this while also quantifying its inherent uncertainty. In Chapter 3, we'll dig into the elements of hypothesis testing, a core method of data analytics.

Exercises

Using Excel and your knowledge of probability, consider the following:

1. What is the expected value of a six-sided die toss?
2. Consider a variable that is normally distributed with a mean of 100 and standard deviation of 10.
   - What is the probability that an observation from this variable would take the value 87?
   - What percentage of observations would you expect to find between 80 and 120?
3. If the expected value of a European roulette spin is 18, does that mean you are better off betting on 18 than other numbers?

Foundations of Inferential Statistics

Chapter 1 provided a framework for exploring a dataset by classifying, summarizing, and visualizing variables. Though this is an essential start to analytics, we usually don't want it to end there: we would like to know whether what we see in our sample data can *generalize* to a larger population.

The thing is, we don't actually know what we'll find in the population, because we don't have the data for all of it. However, using the principles of probability introduced in Chapter 2, we can quantify our uncertainty that what we see in our sample will also be found in the population.

Estimating the values of a population given a sample is known as *inferential statistics* and is carried out by *hypothesis testing*. That framework is the basis of this chapter. You may have studied inferential statistics in school, which could have easily turned you off the subject, seeming incomprehensible and without application. That's why I'll make this chapter as applied as possible, exploring a real-world dataset using Excel.

By the end of the chapter, you will have a handle on this basic framework that powers much of analytics. We'll continue to build on its application in Chapter 4.

Chapter 1 concluded with an exercise on the *housing* dataset, which will be the focus of this chapter. You can find the dataset in the *datasets* folder of the book repository, under the *housing* subfolder. Make a copy of it, add an index column, and convert this dataset into a table named *housing*.

The Framework of Statistical Inference

The ability to infer characteristics of a population based on a sample seems magical, doesn't it? Just like with any magic trick, inferential statistics may look easy to outsiders. But to those within, it's the culmination of a series of finely tuned steps:

0. *Collect a representative sample.* This technically comes *before* the hypothesis test, but is essential for its success. We must be sure that the sample collected is a fair reflection of the population.

1. *State the hypotheses.* First, we will state a *research hypothesis*, or some statement that motivates our analysis and that we think explains something about our population. We will then state a *statistical hypothesis* to test whether the data supports this explanation.

2. *Formulate an analysis plan.* We will then outline what methods we'll use to conduct this test, and what criteria we'll use to evaluate it.

3. *Analyze the data.* Here is where we actually crunch the numbers, and develop the evidence that we'll use to evaluate our test against.

4. *Make a decision.* It's the moment of truth: we will compare the evaluation criteria from step 2 with the actual results from step 3, and conclude whether the evidence supports our statistical hypothesis.

For each of these steps, I'll provide a brief conceptual overview. We'll then immediately apply these concepts to the *housing* dataset.

Collect a Representative Sample

In Chapter 2 you learned that, thanks to the law of large numbers, the average of sample means should get closer to the expected value as the sample size increases. This forms a rule of thumb for what makes an adequate sample size to conduct inferential statistics. We are assuming, however, that we are working with a *representative* sample or a set of observations that fairly reflect the population. If the sample isn't representative, we'd have no standing to assume its sample mean would approach the population mean with more observations.

Ensuring a representative sample is best handled during the conceptualization and collection phases of research; by the time the data is collected, it's hard to walk back any issues related to sampling. There are many approaches to collecting data, but while it's an important part of the analytics workflow, it is outside the scope of this book.

The best time to ensure a representative sample is during the collection of the data itself. If you're working with a preassembled dataset, consider what steps were taken to meet this objective.

Statistical Bias

Collecting an unrepresentative sample is one of the many ways to introduce *bias* into an experiment. You may think of bias in the cultural sense of an inclination or prejudice for or against some thing or person. This is indeed yet another potential source of bias in data analysis. In other words, we say something is statistically biased if it is calculated in some way that is systematically different from the underlying parameter being estimated. Detecting and correcting for bias is one of the central tasks of analytics.

Getting a representative sample of the population prompts the question: what is the target population? This population can be as general or specific as we want. For example, let's say we're interested in exploring the heights and weights of dogs. Our population could be all dogs, or it could be a specific breed. It could be a certain age group or sex of dogs as well. Some target populations may be of more theoretical importance, or logistically easier to sample. Your target population can be anything, but your sample of that population needs to be representative.

At 546 observations, *housing* is likely a large-enough sample for conducting valid inferential statistics. Is it representative, though? Without some understanding of the collection methods or the target population, it's hard to be sure. This data comes from the peer-reviewed *Journal of Applied Econometrics*, so it is trustworthy. Data that you receive at work may not come to you as polished, so it's worth thinking through or inquiring about the collection and sampling procedures.

As for the data's target population, the data's *readme* file, available in the book repository under *datasets → housing*, indicates it comes from home sales in Windsor, Ontario, Canada. That means that housing prices in Windsor may be the best target population; the findings may or may not, for example, transfer to housing prices across Canada or even Ontario. This is also an older dataset, taken from a paper written in the 1990s, so it's not guaranteed that the findings from it would apply to today's housing market, even in Windsor.

State the Hypotheses

With some comfort that our sample data is representative of the population, we can start to think about what exactly we'd like to infer by stating hypotheses. Maybe you've heard rumors about some trend or unusual phenomenon in your data. Maybe something struck you about the data as you began checking it out during EDA. This is your time to speculate on what you *think* you'll find as the result of your analysis. Going to our *housing* example, I think few would disagree that air conditioning is desirable to have in a home. It stands to reason, then, that homes with air conditioning sell for higher prices than those without. This informal statement about the relationship you'll find in the data is called a *research hypothesis*. Another way to state this relationship is that there is an *effect* of air conditioning on sale price. Homes in Windsor are our *population*, and those homes with and without air conditioning are two of its groups or *subpopulations*.

Now, it's great that you have your hypothesis about how air conditioning affects sale prices. It's crucial as an analyst to have a strong intuition and opinion about your work. But as American engineer W. Edwards Deming famously said, "In God we trust. All others must bring data." What we *really* want to know is whether your speculated relationship is actually present in the population. And for this, we'll need to use inferential statistics.

As you've already seen, statistical language is usually different than everyday language. It can feel pedantic at first, but the nuances reveal a lot about how data analysis works. *Statistical hypotheses* are one such example. To test whether the data supports our proposed relationship, we'll provide two statistical hypotheses like the following. Take a look at them now; they'll be explained later:

H0

There is no difference in the average sale price of homes with or without air conditioning.

Ha

There is a difference in the average sale price of homes with or without air conditioning.

By design, these hypotheses are mutually exclusive, so that if one is true, the other must be false. They are also testable and falsifiable, meaning that we can use real-world evidence to measure and contradict them. These are big-idea topics on the philosophy of science that we can't do full credit to here; the main takeaway is that you want to make sure that your hypotheses can actually be tested with your data.

At this point, we need to leave all preconceived notions about the data behind, such as what was speculated in the research hypothesis. We're now assuming *no* effect. After all, why would we? We only have a *sample* of the population's data, so we'll never truly know the population's true value, or parameter. That's why the first hypothesis, or H0, called the *null hypothesis*, is stated so peculiarly.

On the flip side is the *alternative hypothesis*, or Ha. If there's no evidence in the data for the null hypothesis, then the evidence has to be for the alternative, given the way they are stated. That said, we can never say we've *proven* either to be true, because we don't actually know the population's parameter. It could be that the effect we found in the sample is just a fluke, and we wouldn't have actually found it in the population. In fact, measuring the probability of this happening will be a major element of what we do in hypothesis testing.

The results of hypothesis testing don't "prove" either hypothesis to be correct, because the "true" parameter of the population isn't known in the first place.

Formulate an Analysis Plan

Now that we have our statistical hypotheses teed up, it's time to specify the methods used to test the data. The proper statistical test for a given hypothesis depends on a variety of factors, including the type of variables used in the analysis: continuous, categorical, and so on. This is another reason it's a good idea to classify variables during EDA. Specifically, the test we decide to use depends on the types of our independent and dependent variables.

The study of cause and effect drives much of what we do in analytics; we use *independent* and *dependent* variables to model and analyze these relationships. (Remember, though, that because we are dealing with samples, causation is impossible to claim with certainty.) We talked about the idea of experiments in Chapter 2 as repeatable events generating a defined set of random outcomes. We used the example of a die throw as an experiment; most experiments in real life are more complicated. Let's look at an example.

Say we are researchers interested in what contributes to plant growth. A colleague has speculated that watering the plants could have a positive impact. We decide to try it by running experiments. We provide various amounts of water among observations, making sure to record the data. Then we wait a few days and measure the resulting plant growth. We've got two variables in this experiment: watering amount and plant growth. Can you guess which is our independent and which is our dependent variable?

Watering is the *independent variable* because it's what we as researchers control as part of the experiment. Plant growth is the *dependent* because it's what we've hypothesized will change given a change in the independent variable. The independent variable is often recorded first: for example, plants are watered *first*, and *then* they grow.

Independent variables are generally recorded before dependent variables because cause must come before effect.

Given this example, what's the more sensible way to model the relationship between air conditioning and sale price? It stands to reason that air conditioning is installed first, *then* the house is sold. That makes *airco* and *price* our independent and dependent variables, respectively.

Because we are looking to test the effect of a binary independent variable on a continuous dependent variable, we'll use something called the *independent samples t-test*. Don't worry about memorizing the best test to use for any given occasion. Instead, the objective here is picking up the common framework of making inferences about a population given a sample.

Most statistical tests make some assumptions about their data. If these assumptions aren't met, then test results may be inaccurate. For example, the independent samples t-test assumes that no observation influences another and that each observation is found in one and only one group (that is, they are *independent*). To adequately estimate the population mean, the test generally assumes normally distributed samples; that said, it's possible to circumvent that constraint for larger datasets given the magic of the central limit theorem. Excel will help us bypass another assumption: that the variance of each population is equal.

We know what test we'll use, but we still need to set some rules for implementing it. For one, we need to decide the *statistical significance* of the test. Let's go back to the scenario mentioned previously, where the effect inferred in the sample is just a fluke and wouldn't have been found in the population. This scenario will happen eventually, because we'll never actually know the population mean. In other words, we are *uncertain* about this outcome…and, as you learned in Chapter 2, it's possible to *quantify* uncertainty as a number between 0 and 1. This number is called *alpha* and represents the statistical significance of the test.

The alpha shows how comfortable we are with the possibility that there's really no effect in the populations, but that we found one in the samples due to chance. A common threshold for alpha that we'll use throughout the book is 5%. In other words, we are comfortable with claiming a relationship in the data when there really is none 5% or less of the time.

 This book follows the standard convention of conducting two-tailed hypothesis tests at the 5% statistical significance level.

Other common significance levels include 10% or 1%. There is no one "right" level of alpha; setting it depends on a variety of factors such as the research objective, ease of interpretability, and so forth.

You may be wondering why we would be comfortable *at all* with the possibility of claiming an effect when there is none. In other words, why not an alpha of 0? In this case, we're unable to claim *anything* about our population given the sample. Effectively, with an alpha of 0 we'd be saying that, because we don't ever want to be wrong about the population's true value, it could be anything. To make any inference at all, being wrong is a risk we must take.

We also need to state which *direction* in effect we're interested in. For example, we're assuming a *positive* effect of air conditioning on sale price: that is, the average sale price of homes with air conditioning is greater than those without. However, it could turn out that there's a negative effect: it might be you're dealing with a population that would prefer not to have air conditioning. Or, maybe it's a climate where using air conditioning is rarely justified, and having the unit is a needless expense. These scenarios are theoretically possible; if there's any doubt, then the statistical test should examine for both positive *and* negative effects. This is known as a *two-tailed* (or two-tail, as Excel refers to it) test, and we'll be using it in this book. One-tailed tests are possible, but relatively rare and outside our scope.

This may seem like a lot of windup when we haven't even touched the data yet. But these steps exist to ensure that we as analysts come to the data fairly when we finally make the calculations. The results of our hypothesis test depend on the level of statistical significance and the number of tails tested. As you'll see later, it's very possible that slightly different inputs to the test, such as a different statistical significance level, result in different findings. This offers a real temptation to run the numbers, *then* decide on a specific test for a favorable outcome. However, we want to avoid the urge to engineer the results to fit our agenda.

Analyze the Data

And now, the moment you've likely been waiting for: time to crunch the data. This part of the work often gets the most attention and is what we'll focus on here, but it's worth keeping in mind that it's just one of the many steps of hypothesis testing. Remember, data analysis is an iterative process. It's unlikely (and unwise) that you've performed *no* analysis of this data before conducting a hypothesis test. In fact, exploratory data analysis is designed as a precursor to hypothesis testing, or *confirmatory* data analysis. You should *always* be comfortable with the dataset's descriptive statistics before looking to make inferences about it. In that spirit, let's do that here with our *housing* dataset, then move to the analysis.

Figure 3-1 calculates the descriptive statistics for and visualizes the distributions of *price* for both levels of *airco*. If you need a refresher on how to do this, check out Chapter 1. I relabeled the ToolPak output to help indicate what is being measured in each group.

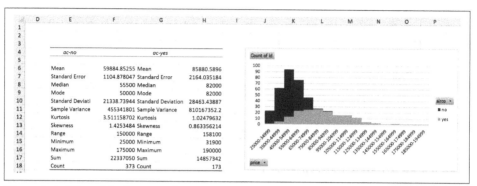

Figure 3-1. EDA of the housing dataset

The histogram in this output shows us that both groups of data are approximately normally distributed, and the descriptive statistics tell us that we have relatively large sample sizes. While far more homes do not have air conditioning (373 without versus 173 with), this is not necessarily a problem for the t-test.

The independent samples t-test is not sensitive to differences in sample sizes between the two groups, so long as each group is sufficiently large. Other statistical tests may be affected by this difference.

Figure 3-1 also gives us the sample means of our groups: approximately $86,000 for homes with air conditioning, and $60,000 for those without. That's good to know, but we'd *really* like to find out whether we could expect such an effect in the population at large. That's where the t-test comes in, and we'll lean yet again on PivotTables and the Data Analysis ToolPak to conduct it.

Insert a PivotTable into a new worksheet, placing *id* in the Rows area, *airco* in the Columns area, and "Sum of price" in the Values area. Clear out all grand totals from the report. This data arrangement will be easy to input into the t-test menu, which can be accessed from the ribbon via Data → Data Analysis → t-Test: Two-Sample Assuming Unequal Variances. The "variances" referred to here are our subpopulation variances. We really don't know whether these are equal, so it's better to choose this option, assuming equal variances for more conservative results.

A dialog box will appear; fill it out as in Figure 3-2. Make sure that the box next to Labels is checked. Immediately above this selection is an option titled Hypothesized

Mean Difference. By default, this is left blank, which means we are testing for a difference of zero. This is precisely our null hypothesis, so we don't have to change anything. Immediately below that line is an option titled Alpha. This is our stated level of statistical significance; Excel defaults to 5%, which is what we want.

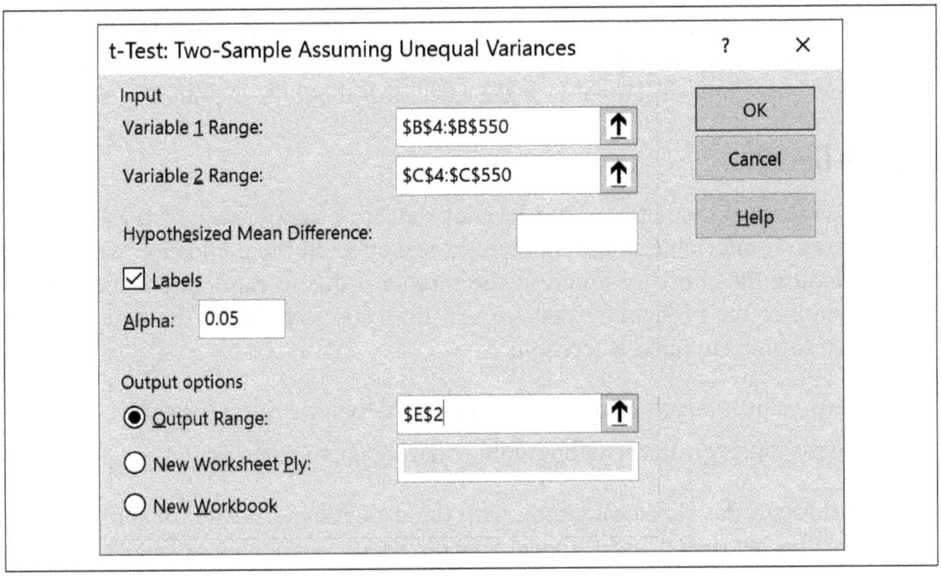

Figure 3-2. t-test setup menu in the ToolPak

The results are shown in Figure 3-3. I've again relabeled each group as *ac-no* and *ac-yes* to clarify what groups are represented. We'll step through selected parts of the output next.

D	E	F	G
1			
2	t-Test: Two-Sample Assuming Unequal Variances		
3			
4		*ac-no*	*ac-yes*
5	Mean	59884.85255	85880.5896
6	Variance	455341801	810167352.2
7	Observations	373	173
8	Hypothesized Mean Difference	0	
9	df	265	
10	t Stat	-10.69882732	
11	P(T<=t) one-tail	9.6667E-23	
12	t Critical one-tail	1.650623976	
13	P(T<=t) two-tail	1.93334E-22	
14	t Critical two-tail	1.968956281	
15			

Figure 3-3. t-test output

First, we're given some information about our two samples in F5:G7: their means, variances, and sample sizes. Our hypothesized mean difference of zero is also included.

We'll skip a few statistics here and focus next on cell F13, $P(T < = t)$ *two-tail*. This probably doesn't make much sense to you, but *two-tail* should sound familiar, as it's the type of test we decided to focus on earlier instead of a one-tail test. This figure is called the *p-value*, and we'll use it to make a decision about the hypothesis test.

Make a Decision

Earlier you learned that alpha is our level of statistical significance, or the level with which we are comfortable assuming there's a real effect in the population when there is not, because the effect we found in the sample is due to random chance. The p-value quantifies the probability that we will find this scenario in the data, and we compare it to alpha to make a decision:

- If the p-value is less than or equal to our alpha, then we reject the null.
- If the p-value is greater than our alpha, then we fail to reject the null.

Let's cut through this statistical jargon with the data at hand. As a probability, the p-value is always between 0 and 1. Our p-value in F13 is very small, so small that Excel has labeled it in scientific notation. The way to read this output is as 1.93 times 10 to the negative 22nd power—a *very* small number. We are saying, then, that if there were really no effect in the population, we'd expect to find the effect that we did in the samples well under less than 1% of the time. This is well below our alpha of 5%, so we can reject the null. When a p-value is so small that scientific notation is required to report it, you'll often see the results simply summarized as "$p < 0.05$."

On the other hand, let's say our p-value was 0.08 or 0.24. In these cases, we would *fail* to reject the null. Why this odd language? Why don't we just say that we "proved" either the null or the alternative? It all goes back to the inherent uncertainty of inferential statistics. We don't ever know the true subpopulation values, so it's safer to start on the premise that they are both equal. The results of our test can confirm or deny evidence for either, but they can never definitively *prove* it.

While p-values are used to make a decision about a hypothesis test, it's also important to know what they *cannot* tell us. For example, a frequent misinterpretation is that the p-value is the probability of making a mistake. In fact, a p-value *assumes* that our null hypothesis is true, regardless of what is found in the sample; the idea of there being a "mistake" in the sample doesn't change this assumption at all. The p-value *only* tells us what percentage of the time we'd find the effect that we did in our sample, given no effect in the population.

The p-value is *not* the probability of making a mistake; rather, it is the probability of finding the effect in the sample that we did, given no effect in the population.

Another common misinterpretation is that the smaller the p-value, the larger the effect. The p-value, however, is only a measure of *statistical* significance: it tells us how *likely* an effect in the population is. The p-value does not indicate the *substantive* significance, or how large that effect size is likely to be. It's common for statistical software to report only statistical and not substantive significance. Our Excel output is one such case: while it returns the p-value, it does not return the *confidence interval*, or the range within which we'd expect to find our population.

We can use the so-called critical value of our test, displayed in cell F14 of Figure 3-3, to derive the confidence interval. The number (1.97) may seem arbitrary, but you can actually make sense of it given what you learned in Chapter 2. With this t-test, we have taken a sample difference in average home prices. Were we to continue taking random samples and plotting the distribution of the mean differences, this distribution would be…that's right, *normal*, due to the central limit theorem.

The Normal Distribution and the T-distribution

For smaller sample sizes, the *t-distribution* is used to derive critical values for a t-test. But as sample sizes increase, critical values converge to those found in the normal distribution. When I refer to specific critical values in this book, I am using those found from the normal distribution; these may vary slightly from what you see in Excel due to sample size. For sample sizes in the hundreds like we've generally been using here, the differences should be negligible

---

With a normal distribution, we can expect about 95% of observations to fall within two standard deviations of the mean thanks to the empirical rule. In the special case of a normally distributed variable with a mean of 0 and standard deviation of 1 (called a *standard* normal distribution), we could say that about 95% of all observations would fall between –2 and 2. To be a little more specific, they would be between –1.96 and 1.96, and that is how the two-tailed critical value is derived. Figure 3-4 illustrates the area within which we'd expect to find the population's parameter with 95% confidence.

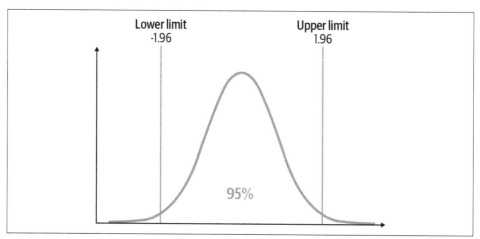

Figure 3-4. The 95% confidence interval and critical value visualized on a histogram

The Test Statistic and the Critical Value

Cell F10 of Figure 3-3 returns the *test statistic*. Though we've been using p-values to make a decision about our hypothesis test, we could also use the test statistic: if it falls outside the inner range of our critical values, we reject the null. The test statistic and the p-value will fundamentally be in agreement; if one indicates significance, so will the other. Because the p-value is often easier to interpret, it's more commonly reported than the test statistic.

Equation 3-1 displays the formula for finding the confidence interval of a two-tailed independent samples t-test. We'll calculate the labeled elements in Excel.

Equation 3-1. Formula for finding the confidence interval

$$c.\ i. = \left(\bar{X}_1 - \bar{X}_2\right) \pm ta_{/2} \times \sqrt{\frac{s_1^2}{n_1} + \frac{s_2^2}{n_2}}$$

To break this formula down, $\left(\bar{X}_1 - \bar{X}_2\right)$ is the point estimate, $ta_{/2}$ is the critical value, and $\sqrt{\frac{s_1^2}{n_1} + \frac{s_2^2}{n_2}}$ is the standard error. The product of the critical value and the standard error is the margin of error.

This equation can be pretty intimidating, so to make it more concrete I've already calculated the confidence interval and its various elements for our example in

Figure 3-5. Rather than get hung up over the formal equation, our focus will be on computing the results and understanding what they tell us.

D	E	F	G
1			
2	t-Test: Two-Sample Assuming Unequal Variances		
3			
4		ac-no	ac-yes
5	Mean	59884.85255	85880.5896
6	Variance	455341801	810167352.2
7	Observations	373	173
8	Hypothesized Mean Difference	0	
9	df	265	
10	t Stat	-10.69882732	
11	P(T<=t) one-tail	9.6667E-23	
12	t Critical one-tail	1.650623976	
13	P(T<=t) two-tail	1.93334E-22	
14	t Critical two-tail	1.968956281	
15			
16	point estimate	25995.73705 =G5-F5	
17	critical value	1.968956281 =F14	
18	standard error	2429.774429 =SQRT((F6/F7)+(G6/G7))	
19	margin of error	4784.119625 =F17*F18	
20	confidence interval lower bound	21211.61742 =F16-F19	
21	confidence interval upper bound	30779.85667 =F16+F19	

Figure 3-5. Calculating the confidence interval in Excel

First, the *point estimate* in cell F16, or the effect in the population we are most likely to find. This is the difference in our sample means. After all, if our sample is representative of the population, the difference in sample and population means should be negligible. But it probably won't be exactly the same; we will derive a range of values within which we are 95% confident we'll find that true difference.

Next, the critical value in cell F17. Excel provided this number for us, but I've included it here for ease of analysis. As described previously, we can use this value to help us find the 95% of values that fall within approximately two standard deviations of the mean.

Now we have the standard error in cell F18. You've actually seen this term before, in the ToolPak's descriptive statistics output; see Figure 3-1 as an example. To understand how the standard error works, imagine if you were to go out and resample housing prices from the population over and over. Each time, you'd get slightly different sample means. That variability is known as the *standard error*. A larger standard error means that a sample is less accurate in representing a population.

The standard error for one sample can be found by dividing its standard deviation by its sample size. Because we are finding the standard error for a *difference* in means, the formula is a bit more complicated, but the same pattern holds: the variability of the samples goes in the numerator, and the number of observations goes in the denominator. This makes sense: we'd expect more variability in a sample difference when each sample mean itself contains more variability; as we collect larger sample sizes, we'd expect them to exhibit less variability from the population.

We will now take the product of the critical value and the standard error to get the *margin of error* in cell F19. This may be a term you've heard of: polls are often reported with this figure included. The margin of error provides an estimate of just how much variability there is around our point estimate. In the case of Figure 3-5, we're saying that while we think the population difference is $25,996, we could be off by as much as $4,784.

Because this is a two-tailed test, this difference could be found in *either direction*. So we'll need to both subtract and add the margin of error to derive the lower and upper bounds of our confidence interval, respectively. Those figures are found in F20 and F21, respectively. The bottom line? With 95% confidence, we believe that the average price of a home with air conditioning has a sale price of between $21,211 and $30,780 higher than one without air conditioning.

Why go through all the trouble to derive a confidence interval? As a measure of substantive rather than statistical significance, it often plays better with general audiences because it translates the results of the statistical hypothesis test back into the language of the research hypothesis. For example, imagine you were a research analyst at a bank reporting the results of this study on home prices to management. These managers wouldn't know where to start conducting a t-test if their careers depended on it—but their careers *do* depend on making smart decisions from that analysis, so you want to make it as intelligible as possible. Which statement do you think will be more helpful?

- "We rejected the null that there is no difference in the average sale price of homes with or without air conditioning at $p < 0.05$."
- "With 95% confidence, the average sale price of homes with air conditioning is approximately between $21,200 and $30,800 higher than those without."

Nearly anyone can make sense of the second statement, whereas the first requires a fair amount of statistical know-how. But confidence intervals aren't just for laypeople: there's also been a push in research and data circles to report them along with p-values. After all, the p-value *only* measures statistical effect, *not* substantive.

But while p-values and confidence intervals show the results from different angles, they are fundamentally *always in agreement*. Let's illustrate this concept by conducting another hypothesis test on the *housing* dataset. This time, we would like to know if there is a significant difference in the average lot size (*lotsize*) of homes with and without a full, finished basement (*fullbase*). This relationship can also be tested with a t-test; I will follow the same steps as before in a new worksheet, which results in

Figure 3-6. (Don't forget to explore the descriptive statistics of these new variables first.)

	D	E	F	G	H
1					
2		t-Test: Two-Sample Assuming Unequal Variances			
3					
4			fullbase-no	fullbase-yes	
5		Mean	5074.814085	5290.502618	
6		Variance	4683966.27	4726820.23	
7		Observations	355	191	
8		Hypothesized Mean Difference	0		
9		df	387		
10		t Stat	-1.107303893		
11		P(T<=t) one-tail	0.134425163		
12		t Critical one-tail	1.648800515		
13		P(T<=t) two-tail	0.268850325		
14		t Critical two-tail	1.966112774		
15					
16		point estimate	215.6885333	=G5-F5	
17		critical value	1.966112774	=F14	
18		standard error	194.7871173	=SQRT((F6/F7)+(G6/G7))	
19		margin of error	382.9734396	=F17*F18	
20		confidence interval lower bound	-167.2849064	=F16-F19	
21		confidence interval upper bound	598.6619729	=F16+F19	
22					

Figure 3-6. The effect on lot size of having a full finished basement

The results of this test are *not* statistically significant: based on the p-value of 0.27, we'd expect to find the effect we did in over one-quarter of our samples, assuming no effect in the population. As for the substantive significance, we are 95% confident that the difference in average lot size is between approximately 167 square feet less and 599 square feet more. In other words, the true difference could be positive *or* negative, we can't say for sure. Based on either of these results, we fail to reject the null: there does not appear to be a significant difference in average lot size. These results will always agree because they are both based in part on the level of statistical significance: alpha determines how we evaluate the p-value and sets the critical value used to derive the confidence interval.

Hypothesis Testing and Data Mining

Whether we really *did* expect to find a significant relationship between lot size and a full, finished basement is questionable. After all, how these variables are related is less clear than how the presence of air conditioning might affect sale price. In fact, I chose to test this relationship only to get an intended *insignificant* relationship for the purposes of demonstration. In most other cases, it would be more tempting to mine the data and look for *significant* relationships. Cheap computing has allowed for a more freewheeling approach to data analysis, but if you can't explain the results of your analysis based on logic, theory, or prior empirical evidence, you should treat it cautiously—no matter how strong or powerful the findings.

If you've ever built a financial model, you might be familiar with doing a what-if analysis on your work to see how its output might change given your inputs or assumptions. In that same spirit, let's examine how the results of our basement/lot size t-test might have been different. Because we'll be manipulating the results of our ToolPak output, it's wise to copy and paste the data in cells E2:G21 into a new range so the original is preserved. I'll place mine in cells J2:L22 of the current worksheet. I will also relabel my output and highlight the cells K6:L6 and K14 so that it's clear they've been tampered with.

Let's manipulate the sample sizes and critical value here. Without looking at the resulting confidence interval, try to guess what will happen based on what you know about how these figures relate. First, I will set the sample size to 550 observations for each group. This is a dangerous game to play; we didn't *actually* collect 550 observations, but to understand statistics you sometimes have to get your hands dirty. Next, we'll change our statistical significance from 95% to 90%. The resulting critical value is 1.64. This is also dicey; statistical significance should be locked in prior to analysis for the reason that you are about to see right now.

Figure 3-7 displays the result of this what-if analysis. Our confidence interval of between \$1 and \$430 indicates statistical significance, although just barely—it's awfully close to zero.

D I	J	K	L
1			
2	t-Test: Two-Sample Assuming Unequal Variances WHAT-IF ANALYSIS		
3			
4		fullbase-no	fullbase-yes
5	Mean	5074.814085	5290.502618
6	Variance	4683966.27	4726820.23
7	Observations	550	550
8	Hypothesized Mean Difference	0	
9	df	387	
10	t Stat	-1.107303893	
11	P(T<=t) one-tail	0.134425163	
12	t Critical one-tail	1.648800515	
13	P(T<=t) two-tail	0.268850325	
14	t Critical two-tail	1.64	
15			
16	point estimate	215.6885333 =L5-K5	
17	critical value	1.64 =K14	
18	standard error	130.8071898 =SQRT((K6/K7)+(L6/L7))	
19	margin of error	\$215 =K17*K18	
20	confidence interval lower bound	\$1 =K16-K19	
21	confidence interval upper bound	\$430 =K16+K19	
22			

Figure 3-7. A what-if analysis of the confidence interval

There are ways to calculate the corresponding p-value, but because you know it is always fundamentally in agreement with the confidence interval, we'll skip that exercise. Our test is now significant, and that can make all the difference for funding, fame, and glory.

The moral of the story is that the results of hypothesis testing can be easily gamed. Sometimes, all it takes is a different level of statistical significance to tip the balance to rejecting the null. Resampling or, in our example, falsely bulking up the number of observations could also do it. Even absent foul play, there will always be a gray area in claiming to find the parameter of a population you don't actually know.

It's Your World…the Data's Only Living in It

It's tempting to go into autopilot when conducting inferential statistics, just plugging and chugging p-values without regard for broader considerations about data collection or substantive significance. You've already seen how sensitive the results can be to changes in statistical significance or sample size. To show another possibility, let's take one more example from the *housing* dataset.

On your own, test for a significant difference in the sale price of homes with and without gas for water heating. The relevant variables are *price* and *gashw*. The results are shown in Figure 3-8.

	F	G	H	I	J
1					
2					
3		t-Test: Two-Sample Assuming Unequal Variances			
4					
5			gas-yes	gas-no	
6		Mean	79428	67579.06334	
7		Variance	923472100	698250450.3	
8		Observations	25	521	
9		Hypothesized Mean Difference	0		
10		df	26		
11		t Stat	1.915131244		
12		P(T<=t) one-tail	0.033268787		
13		t Critical one-tail	1.70561792		
14		P(T<=t) two-tail	0.066537575		
15		t Critical two-tail	2.055529439		
16					
17		point estimate	-11848.93666	=I6-H6	
18		critical value	2.055529439	=H15	
19		standard error	6187.010263	=SQRT((H7/H8)+ (I7/I8))	
20		margin of error	12717.58173	=H18*H19	
21		confidence interval lower bound	-24566.51839	=H17-H20	
22		confidence interval upper bound	868.6450729	=H17+H20	
23					

Figure 3-8. t-test results of the effect on sale price of using gas

Going by the p-value alone, we should fail to reject the null: after all, it's greater than 0.05. But 0.067 is not *that* different, so it's worth paying closer attention here. For one thing, consider the sample sizes: at just 25 observations of homes with gas, it may be worth collecting more data before definitively rejecting the null. Granted, you probably would have observed this sample size prior to running the test, during descriptive statistics.

By the same token, the confidence interval posits that the true difference is anywhere between approximately $900 less and $24,500 more. With that kind of money on the table, it's worth digging further into the problem. If you were just to blindly reject the null due to the p-value, you may miss out on a potentially important relationship. Be aware of these potential "edge cases": if one already came up here in this dataset, you can bet you'll find more in your data work.

 Statistics and analytics are powerful tools for making sense of the world, but they're just that: tools. Without a skilled craftsperson in control, they can be useless at best and harmful at worst. Don't be content to take the p-value on its face; consider the broader context of how statistics works and the objective you're aiming to meet (without gaming the results, as you've seen is possible). Remember: it's your world, the data's only living in it.

Conclusion

You may have been wondering earlier why, in a book on analytics, we spent a chapter on the seemingly obscure topic of probability. I hope the connection is now clear: because we don't know parameters of the population, we must quantify this uncertainty as a probability. In this chapter, we've used the framework of inferential statistics and hypothesis testing to explore the difference in means between two groups. In the next, we'll use it to examine the influence of one continuous variable on another, in a method you may have heard of: linear regression. Although a different test, the statistical framework behind it remains the same.

Exercises

Now it's your turn to make probabilistic inferences about a dataset. Find the *tips.xlsx* dataset in the *datasets* folder and *tips* subfolder of the book's companion repository (*https://oreil.ly/1hlYj*), and try the following exercises:

1. Test the relationship between the time of day (lunch or dinner) and the total bill:

   - What are your statistical hypotheses?

   - Are your results statistically significant? What evidence does this lend to your hypotheses?

   - What is the estimated effect size?

2. Answer the same questions, but for the relationship between the time of day and the tip.

## CHAPTER 4

Correlation and Regression

Have you heard that ice cream consumption is linked to shark attacks? Apparently Jaws has a lethal appetite for mint chocolate chip. Figure 4-1 visualizes this proposed relationship.

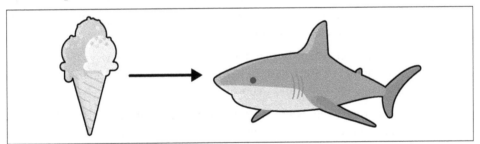

Figure 4-1. The proposed relationship between ice cream consumption and shark attacks

"Not so," you may retort. "This does not necessarily mean that shark attacks are triggered by ice cream consumption."

"It could be," you reason, "that as the outside temperature increases, more ice cream is consumed. People also spend more time near the ocean when the weather is warm, and *that* coincidence leads to more shark attacks."

"Correlation Does Not Imply Causation"

You've likely heard repeatedly that "correlation does not imply causation."

In Chapter 3, you learned that *causation* is a fraught expression in statistics. We really only reject the null hypothesis because we simply don't have all the data to claim causality for sure. That semantic difference aside, does correlation have *anything* to do with causation? The standard expression somewhat oversimplifies their relationship;

you'll see why in this chapter using the tools of inferential statistics you picked up earlier.

This will be our final chapter conducted primarily in Excel. After that, you will have sufficiently grasped the framework of analytics to be ready to be move into R and Python.

Introducing Correlation

So far, we've mostly been analyzing statistics one variable at a time: we've found the average reading score or the variance in housing prices, for example. This is known as *univariate* analysis.

We've also done a bit of *bivariate* analysis. For example, we compared the frequencies of two categorical variables using a two-way frequency table. We also analyzed a continuous variable when grouped by multiple levels of a categorical variable, finding descriptive statistics for each group.

We will now calculate a bivariate measure of two continuous variables using correlation. More specifically, we will use the *Pearson correlation coefficient* to measure the strength of the *linear relationship* between two variables. Without a linear relationship, the Pearson correlation is unsuitable.

So, how do we know our data is linear? There are more rigorous ways to check, but, as usual, a visualization is a great start. In particular, we will use a *scatterplot* to depict all observations based on their x and y coordinates.

If it appears a line could be drawn through the scatterplot that summarizes the overall pattern, then it's a linear relationship and the Pearson correlation can be used. If you would need a curve or some other shape to summarize the pattern, then the opposite holds. Figure 4-2 depicts one linear and two nonlinear relationships.

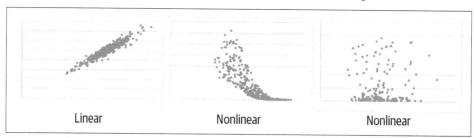

Figure 4-2. Linear versus nonlinear relationships

In particular, Figure 4-2 gives an example of a *positive* linear relationship: as values on the x-axis increase, so do the values on the y-axis (at a linear rate).

It's also possible to have a *negative* correlation, where a negative line summarizes the relationship, or no correlation at all, in which a flat line summarizes it. These different types of linear relationships are shown in Figure 4-3. Remember, these all must be linear relationships for correlation to apply.

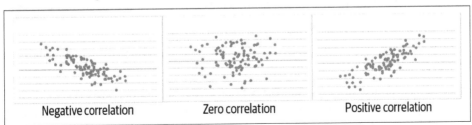

Figure 4-3. Negative, zero, and positive correlations

Once we've established that the data is linear, we can find the correlation coefficient. It always takes a value between –1 and 1, with –1 indicating a perfect negative linear relationship, 1 a perfect positive linear relationship, and 0 no linear relationship at all. Table 4-1 shows some rules of thumb for evaluating the strength of a correlation coefficient. These are not official standards by any means, but a useful jumping-off point for interpretation.

Table 4-1. Interpretation of correlation coefficients

Correlation coefficient	Interpretation
–1.0	Perfect negative linear relationship
–0.7	Strong negative relationship
–0.5	Moderate negative relationship
–0.3	Weak negative relationship
0	No linear relationship
+0.3	Weak positive relationship
+0.5	Moderate positive relationship
+0.7	Strong positive relationship
+1.0	Perfect positive linear relationship

With the basic conceptual framework for correlations in mind, let's do some analysis in Excel. We will be using a vehicle mileage dataset; you can find *mpg.xlsx* in the *mpg* subfolder of the book repository's *datasets* folder (*https://oreil.ly/ygWQn*). This is a new dataset, so take some time to learn about it: what types of variables are we working with? Summarize and visualize them using the tools covered in Chapter 1. To help with subsequent analysis, don't forget to add an index column and convert the dataset into a table, which I will call *mpg*.

Excel includes the CORREL() function to calculate the correlation coefficient between two arrays:

```
CORREL(array1, array2)
```

Let's use this function to find the correlation between weight and mpg in our dataset:

```
=CORREL(mpg[weight], mpg[mpg])
```

This indeed returns a value between –1 and 1: –0.832. (Do you remember how to interpret this?)

A *correlation matrix* presents the correlations across all pairs of variables. Let's build one using the Data Analysis ToolPak. From the ribbon, head to Data → Data Analysis → Correlation.

Remember that this is a measure of linear relationship between two *continuous* variables, so we should exclude categorical variables like *origin* and be judicious about including discrete variables like *cylinders* or *model.year*. The ToolPak insists on all variables being in a contiguous range, so I will cautiously include *cylinders*. Figure 4-4 shows what the ToolPak source menu should look like.

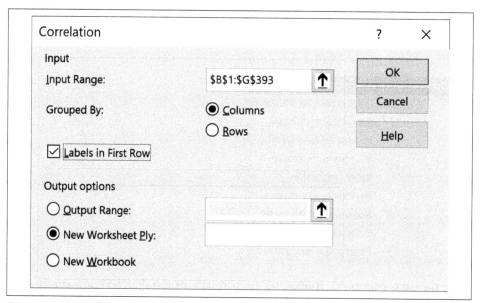

Figure 4-4. Inserting a correlation matrix in Excel

This results in a correlation matrix as shown in Figure 4-5.

⊿	A	B	C	D	E	F	G	H
1		*mpg*	*cylinders*	*displacement*	*horsepower*	*weight*	*acceleration*	
2	mpg	1						
3	cylinders	-0.777617508	1					
4	displacement	-0.805126947	0.950823301	1				
5	horsepower	-0.778426784	0.842983357	0.897257002	1			
6	weight	-0.832244215	0.89752734	0.932994404	0.864537738	1		
7	acceleration	0.423328537	-0.504683379	-0.543800497	-0.68919551	-0.416839202	1	
8								
9								
10								

Figure 4-5. Correlation matrix in Excel

We can see the –0.83 in cell B6: it's the intersection of *weight* and *mpg*. We would also see the same value in cell F2, but Excel left this half of the matrix blank, as it's redundant information. All values along the diagonal are 1, as any variable is perfectly correlated with itself.

 The Pearson correlation coefficient is only a suitable measure when the relationship between the two variables is linear.

We've made a big leap in our assumptions about our variables by analyzing their correlations. Can you think of what that is? *We assumed their relationship is linear.* Let's check that assumption with scatterplots. Unfortunately, there is no way in basic Excel to generate scatterplots of each pair of variables at once. For practice, consider plotting them all, but let's try it with the *weight* and *mpg* variables. Highlight this data, then head to the ribbon and click Insert → Scatter.

I will add a custom chart title and relabel the axes to aid in interpretation. To change the chart title, double-click on it. To relabel the axes, click the perimeter of the chart and then select the plus sign that appears to expand the Chart Elements menu. (On Mac, click inside the chart and then Chart Design → Add Chart Element.) Select Axis Titles from the menu. Figure 4-6 shows the resulting scatterplot. It's not a bad idea to include units of measurement on the axes to help outsiders make sense of the data.

Figure 4-6 looks basically like a negative linear relationship, with a greater spread given lower weights and higher mileages. By default, Excel plotted the first variable in our data selection along the x-axis and the second along the y-axis. But why not the other way around? Try switching the order of these columns in your worksheet so that *weight* is in column E and *mpg* in column F, then insert a new scatterplot.

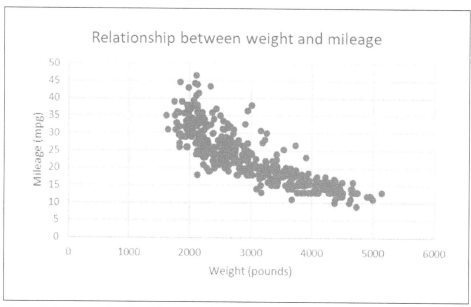

Figure 4-6. Scatterplot of weight and mileage

Figure 4-7 shows a mirror image of the relationship. Excel is a great tool, but as with any tool, you have to tell it what to do. Excel will calculate correlations regardless of whether the relationship is linear. It will also make you a scatterplot without concern for which variable should go on which axis.

So, which scatterplot is "right?" Does it matter? By convention, the independent variable goes on the x-axis, and dependent on the y-axis. Take a moment to consider which is which. If you're not sure, remember that the independent variable is generally the one measured first.

Our independent variable is *weight* because it was determined by the design and building of the car. *mpg* is the *dependent* variable because we assume it's affected by the car's weight. This puts *weight* on the x-axis and *mpg* on the y-axis.

In business analytics, it is uncommon to have collected data just for the sake of statistical analysis; for example, the cars in our *mpg* dataset were built to generate revenue, not for a research study on the impact of weight on mileage. Because there are not always clear independent and dependent variables, we need to be all the more aware of *what* these variables are measuring, and *how* they are measured. This is why having some knowledge of the domain you are studying, or at least descriptions of your variables and how your observations were collected, is so valuable.

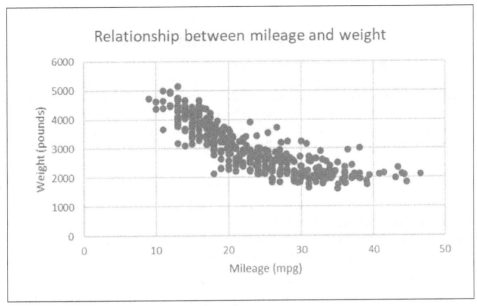

Figure 4-7. Scatterplot of mileage and weight

From Correlation to Regression

Though it's conventional to place the independent variable on the x-axis, it makes no difference to the related correlation coefficient. However, there is a big caveat here, and it relates to the earlier idea of using a line to summarize the relationship found by the scatterplot. This practice begins to diverge from correlation, and it's one you may have heard of: *linear regression.*

Correlation is agnostic to which variable you call independent and which you call dependent; that doesn't factor into its definition as "the extent to which two variables move together linearly."

On the other hand, linear regression is inherently affected by this relationship as "the estimated impact of a unit change of the independent variable X on the dependent variable Y."

You are going to see that the line we fit through our scatterplot can be expressed as an equation; unlike the correlation coefficient, this equation depends on how we define our independent and dependent variables.

Linear Regression and Linear Models

You will often hear linear regression referred to as a linear *model*, which is just one of many statistical models. Just like a model train you might build, a statistical model serves as a workable approximation of a real-life subject. In particular, we use statistical models to understand the relationship between dependent and independent variables. Models won't be able to explain *everything* about what they represent, but that doesn't mean they can't help. But as British mathematician George Box famously stated, "All models are wrong, but some are useful."

Like correlation, linear regression assumes that a linear relationship exists between the two variables. Other assumptions do exist and are important to consider when modeling data. For example, we do not want to have extreme observations that may disproportionately affect the overall trend of the linear relationship.

For the purposes of our demonstration, we will overlook this and other assumptions for now. These assumptions are often difficult to test using Excel; your knowledge of statistical programming will serve you well when looking into the deeper points of linear regression.

Take a deep breath; it's time for another equation:

Equation 4-1. The equation for linear regression

$$Y = \beta_0 + \beta_1 \times X + \epsilon$$

The goal of Equation 4-1 is to predict our dependent variable Y. That's the left side. You may remember from school that a line can be broken into its *intercept* and *slope*. That's where β_0 and $\beta_1 \times X_i$, respectively, come in. In the second term, we multiply our independent variable by a slope *coefficient*.

Finally, there will be a part of the relationship between our independent and dependent variable that can be explained not by the model per se but by some external influence. This is known as the model's *error* and is indicated by ε_i.

Earlier we used the independent samples t-test to examine a significant difference in means between two groups. Here, we are measuring the linear influence of one continuous variable on another. We will do this by examining whether the *slope* of the fit regression line is statistically different than zero. That means our hypothesis test will work something like this:

H0: There is no linear influence of our independent variable on our dependent variable. (The slope of the regression line equals zero.)

Ha: There is a linear influence of our independent variable on our dependent variable. (The slope of the regression line does not equal zero.)

Figure 4-8 shows some examples of what significant and insignificant slopes might look like.

Remember, we don't have *all* the data, so we don't know what the "true" slope would be for the population. Instead, we are inferring whether, given our sample, this slope would be statistically different from zero. We can use the same p-value methodology to estimate the slope's significance that we did to find the difference in means of two groups. We will continue to conduct two-tailed tests at the 95% confidence interval. Let's jump into finding the results using Excel.

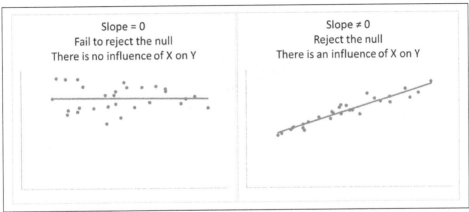

Figure 4-8. Regression models with significant and insignificant slopes

Linear Regression in Excel

In this demo of linear regression on the *mpg* dataset in Excel, we test whether a car's weight (*weight*) has a significant influence on its mileage (*mpg*). That means our hypotheses will be:

H0: There is no linear influence of weight on mileage.

Ha: There is a linear influence of weight on mileage.

Before getting started, it's a good idea to write out the regression equation using the specific variables of interest, which I've done in Equation 4-2:

Equation 4-2. Our regression equation for estimating mileage

$$mpg = \beta_0 + \beta_1 \times weight + \epsilon$$

Let's start with visualizing the results of the regression: we already have the scatterplot from Figure 4-6, now it's just a matter of overlaying or "fitting" the regression line onto it. Click on the perimeter of the plot to launch the "Chart Elements" menu. Click on "Trendline," then "More Options" to the side. Click the radio button at the bottom of the "Format Trendline" screen reading "Display Equation on chart."

Now let's click on the resulting equation on the graph to add bold formatting and increase its font size to 14. We'll make the trendline solid black and give it a 2.5-point width by clicking on it in the graph, then going to the paint bucket icon at the top of the Format Trendline menu. We now have the making of linear regression. Our scatterplot with trendline looks like Figure 4-9. Excel also includes the *regression* equation we are looking for from Equation 4-2 to estimate a car's mileage based on its weight.

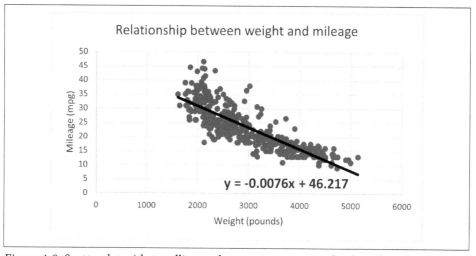

Figure 4-9. Scatterplot with trendline and regression equation for the effect of weight on mileage

We can place the intercept before the slope in our equation to get Equation 4-3.

Equation 4-3. Equation 4-3. Our fit regression equation for estimating mileage

$$mpg = 46.217 - 0.0076 \times weight$$

Notice that Excel does not include the error term as part of the regression equation. Now that we've fit the regression line, we've quantified the difference between what values we expect from the equation and what values are found in the data. This difference is known as the *residual*, and we'll come back to it later in this chapter. First, we'll get back to what we set out to do: establish statistical significance.

It's great that Excel fit the line for us and gave us the resulting equation. But this does *not* give us enough information to conduct the hypothesis test: we still don't know whether the line's slope is statistically different than zero. To get this information, we will again use the Analysis ToolPak. From the ribbon, go to Data → Data Analysis → Regression. You'll be asked to select your Y and X ranges; these are your dependent and independent variables, respectively. Make sure to indicate that your inputs include labels, as shown in Figure 4-10.

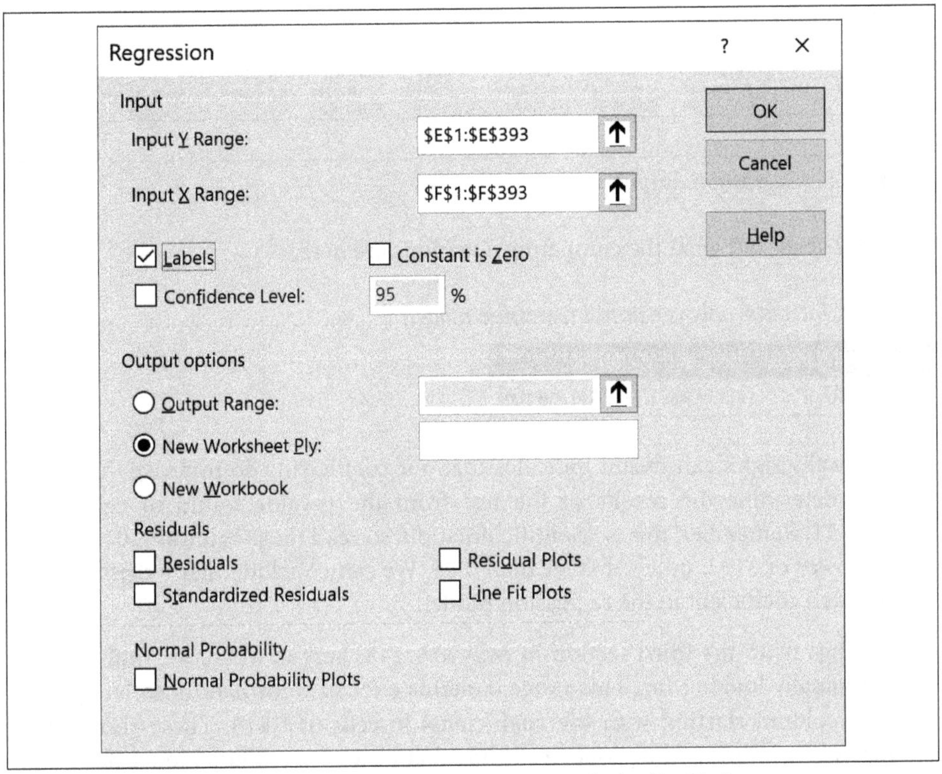

Figure 4-10. Menu settings for deriving a regression with the ToolPak

This results in quite a lot of information, which is shown in Figure 4-11. Let's step through it.

Ignore the first section in cells A3:B8 for now; we will return to it later. Our second section in A10:F14 is labeled ANOVA (short for *analysis of variance*). This tells us whether our regression performs significantly better with the coefficient of the slope included versus one with just the intercept.

	A	B	C	D	E	F	G	H	I
1	SUMMARY OUTPUT								
2									
3	*Regression Statistics*								
4	Multiple R	0.832244215							
5	R Square	0.692630433							
6	Adjusted R Square	0.691842306							
7	Standard Error	4.332712097							
8	Observations	392							
9									
10	ANOVA								
11		*df*	*SS*	*MS*	*F*	*Significance F*			
12	Regression	1	16497.75976	16497.75976	878.8308864	6.0153E-102			
13	Residual	390	7321.233706	18.77239412					
14	Total	391	23818.99347						
15									
16		*Coefficients*	*Standard Error*	*t Stat*	*P-value*	*Lower 95%*	*Upper 95%*	*Lower 95.0%*	*Upper 95.0%*
17	Intercept	46.21652455	0.798672463	57.86668086	1.6231E-193	44.64628231	47.78676679	44.64628231	47.78676679
18	weight	-0.007647343	0.000257963	-29.64508199	6.0153E-102	-0.008154515	-0.00714017	-0.008154515	-0.00714017
19									

Figure 4-11. Regression output

Table 4-2 spells out what the competing equations are here.

Table 4-2. Intercept-only versus full regression model

Incercept-only model	Model with coefficients
$mpg = 46.217$	$mpg = 46.217 - 0.0076 \times weight$

A statistically significant result indicates that our coefficients do improve the model. We can determine the results of the test from the p-value found in cell `F12` of Figure 4-11. Remember, this is scientific notation, so read the p-value as 6.01 times 10 to the power of –102: much smaller than 0.05. We can conclude that *weight* is worth keeping as a coefficient in the regression model.

That brings us to the third section in cells `A16:I18`; here is where we find what we were originally looking for. This range contains a lot of information, so let's go column by column starting with the coefficients in cells `B17:B18`. These should look familiar as the intercept and slope of the line that were given in Equation 4-3.

Next, the standard error in `C17:C18`. We talked about this in Chapter 3: it's a measure of variability across repeated samples and in this case can be thought of as a measure of our coefficients' precision.

We then have what Excel calls the "t Stat," otherwise known as the t-statistic or test statistic, in `D17:D18`; this can be derived by dividing the coefficient by the standard error. We can compare it to our critical value of 1.96 to establish statistical significance at 95% confidence.

It's more common, however, to interpret and report on the p-value, which gives the same information. We have two p-values to interpret. First, the intercept's coefficient in `E17`. This tells us whether the intercept is significantly different than zero. The significance of the intercept is *not* part of our hypothesis test, so this information is irrelevant. (This is another good example of why we can't always take Excel's output at face value.)

 While most statistical packages (including Excel) report the p-value of the intercept, it's usually not relevant information.

Instead, we want the p-value of *weight* in cell E18: this is related to the line's slope. The p-value is well under 0.05, so we reject the null and conclude that weight does likely influence mileage. In other words, the line's slope is significantly different than zero. Just like with our earlier hypothesis tests, we will shy away from concluding that we've "proven" a relationship, or that more weight *causes* lower mileage. Again, we are making inferences about a population based on a sample, so uncertainty is inherent.

The output also gives us the 95% confidence interval for our intercept and slope in cells F17:I18. By default, this is stated twice: had we asked for a different confidence interval in the input menu, we'd have received both here.

Now that you're getting the hang of interpreting the regression output, let's try making a *point estimate* based on the equation line: what would we expect the mileage to be for a car weighing 3,021 pounds? Let's plug it into our regression equation in Equation 4-4:

Equation 4-4. Equation 4-4. Making a point estimate based on our equation

$$mpg = 46.217 - 0.0076 \times 3021$$

Based on Equation 4-4, we expect a car weighing 3,021 pounds to get 23.26 miles per gallon. Take a look at the source dataset: there *is* an observation weighing 3,021 pounds (Ford Maverick, row 101 in the dataset) and it gets 18 miles per gallon, not 23.26. *What gives?*

This discrepancy is the *residual* that was mentioned earlier: it's the difference between the values we estimated in the regression equation and those that are found in the actual data. I've included this and some other observations in Figure 4-12. The scatterpoints represent what values are actually found in the dataset, and the line represents what values we predicted with the regression.

It stands to reason that we'd be motivated to minimize the difference between these values. Excel and most regression applications use *ordinary least squares* (OLS) to do this. Our goal in OLS is to minimize residuals, specifically, the *sum of squared residuals*, so that both negative and positive residuals are measured equally. The lower the sum of squared residuals, the less of a difference there is between our actual and expected values, and the better our regression equation is at making estimates.

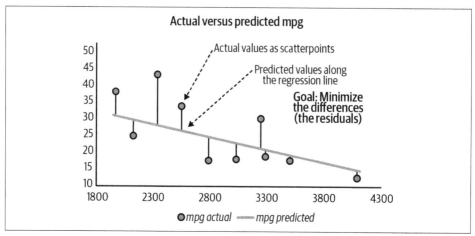

Figure 4-12. Residuals as the differences between actual and predicted values

We learned from the p-value of our slope that there is a significant relationship between independent and dependent variables. But this does not tell us how *much* of the variability in our dependent variable is explained by our independent variable.

Remember that variability is at the heart of what we study as analysts; variables vary, and we want to study *why* they vary. Experiments let us do that, by understanding the relationship between an independent and dependent variable. But we won't be able to explain everything about our dependent variable with our independent variable. There will always be some unexplained error.

R-squared, or the coefficient of determination (which Excel refers to as *R-square*), expresses as a percentage how much variability in the dependent variable is explained by our regression model. For example, an R-squared of 0.4 indicates that 40% of variability in Y can be explained by the model. This means that 1 minus R-squared is what variability *can't* be explained by the model. If R-squared is 0.4, then 60% of Y's variability is unaccounted for.

Excel calculates R-squared for us in the first box of regression output; take a look back to cell B5 in Figure 4-11. The square root of R-squared is multiple R, which is also seen in cell B4 of the output. Adjusted R-square (cell B6) is used as a more conservative estimate of R-squared for a model with multiple independent variables. This measure is of interest when conducting *multiple* linear regression, which is beyond the scope of this book.

Multiple Linear Regression

This chapter has focused on *univariate* linear regression, or the influence of one independent on one dependent variable. It's also possible to build multiple, or *multivariate*, regression models to estimate the influence of several independent variables on a dependent variable. These independent variables can include categorical, not just continuous, variables, interactions between variables, and more. For a detailed look at performing more complex linear regression in Excel, check out Conrad Carlberg's *Regression Analysis Microsoft Excel* (Que).

There are other ways than R-squared to measure the performance of regression: Excel includes one of them, the standard error of the regression, in its output (cell B7 in Figure 4-11). This measure tells us the average distance that observed values deviate from the regression line. Some analysts prefer this or other measures to R-squared for evaluating regression models, although R-squared remains a dominant choice. Regardless of preferences, the best evaluation often comes from evaluating multiple figures in their proper context, so there's no need to swear by or swear off any one measure.

Congratulations: you conducted and interpreted a complete regression analysis.

Rethinking Our Results: Spurious Relationships

Based on their temporal ordering and our own logic, it's nearly absolute in our mileage example that *weight* should be the independent variable and *mpg* the dependent. But what happens if we fit the regression line with these variables reversed? Go ahead and give it a try using the ToolPak. The resulting regression equation is shown in Equation 4-5.

> *Equation 4-5. Equation 4-5. A regression equation to estimate weight based on mileage*
>
> $weight = 5101.1 - 90.571 \times mpg$

We can flip our independent and dependent variables and get the same correlation coefficient. But when we change them for regression, *our coefficients change*.

Were we to find out that *mpg* and *weight* were both influenced simultaneously by some outside variable, then neither of these models would be correct. And this is the same scenario that we're faced with in ice cream consumption and shark attacks. It's silly to say that ice cream consumption has any influence on shark attacks, because both of these are influenced by temperature, as Figure 4-13 depicts.

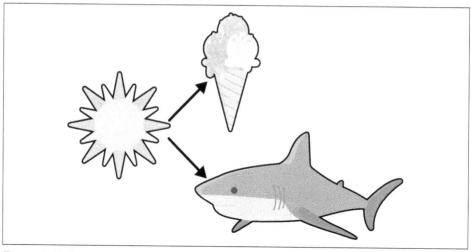

Figure 4-13. Ice cream consumption and shark attacks: a spurious relationship

This is called a *spurious* relationship. It's frequently found in data, and it may not be as obvious as this example. Having some domain knowledge of the data you are studying can be invaluable for detecting spurious relationships.

 Variables can be correlated; there could even be evidence of a causal relationship. But the relationship might be driven by some variable you've not even accounted for.

Conclusion

Remember this old phrase?

> Correlation doesn't imply causation.

Analytics is highly incremental: we usually layer one concept on top of the next to build increasingly complex analyses. For example, we'll always start with descriptive statistics of the sample before attemping to infer parameters of the population. While correlation may not imply causation, causation is built on the foundations of correlation. That means a better way to summarize the relationship might be:

> Correlation is a necessary but not sufficient condition for causation.

We've just scratched the surface of inferential statistics in this and previous chapters. A whole world of tests exists, but all of them use the same framework of *hypothesis testing* that we've used here. Get this process down, and you'll be able to test for all sorts of different data relationships.

Advancing into Programming

I hope you've seen and agree that Excel is a fantastic tool for learning statistics and analytics. You got a hands-on look at the statistical principles that power much of this work, and learned how to explore and test for relationships in real datasets.

That said, Excel can have diminishing returns when it comes to more advanced analytics. For example, we've been checking for properties like normality and linearity using visualizations; this is a good start, but there are more robust ways to test them (often, in fact, using statistical inference). These techniques often rely on matrix algebra and other computationally intensive operations that can be tedious to derive in Excel. While add-ins are available to make up for these shortcomings, they can be expensive and lack particular features. On the other hand, as open source tools R and Python are free, and include many app-like features called *packages* that serve nearly any use case. This environment will allow you to focus on the conceptual analysis of your data rather than raw computation, but you will need to learn how to program. These tools, and the analytics toolkit in general, will be the focus of Chapter 5.

Exercises

Practice your correlation and regression chops by analyzing the *ais* dataset found in the book repository's *datasets* folder (*https://oreil.ly/hazKQ*). This dataset includes height, weight, and blood readings from male and female Australian athletes of different sports.

With the dataset, try the following:

1. Produce a correlation matrix of the relevant variables in this dataset.

2. Visualize the relationship of *ht* and *wt*. Is this a linear relationship? If so, it is negative or positive?

3. Of *ht* and *wt*, which do you presume is the independent and dependent variable?

   - Is there a significant influence of the independent variable on the dependent variable?

   - What is the slope of your fit regression line?

   - What percentage of the variance in the dependent variable is explained by the independent variable?

4. This dataset contains a variable for body mass index, *bmi*. If you are not familiar with this metric, take a moment to research how it's calculated. Knowing this, would you want to analyze the relationship between *ht* and *bmi*? Don't hesitate to lean on common sense here rather than just statistical reasoning.

## CHAPTER 5

The Data Analytics Stack

By this point in the book, you are well versed in the key principles and methods of analytics, having learned them in Excel. This chapter serves as an interlude to the following parts of the book, where you'll pivot that existing knowledge into using R and Python.

This chapter will further delineate the disciplines of statistics, data analytics, and data science, and we'll take a deep dive into how Excel, R, and Python all play into what I call the *data analytics stack*.

Statistics Versus Data Analytics Versus Data Science

The focus of this book is helping you master principles of data analytics. But as you've seen, statistics is so core to analytics that it's often hard to delineate where one field ends and the other begins. To compound the confusion, you may also be interested in how data science fits into the mix. Let's take a moment to tighten these distinctions.

Statistics

Statistics is foremost concerned with the methods for collecting, analyzing, and presenting data. We've borrowed a *lot* from the field: for example, we made inferences about a population given a sample, and we depicted distributions and relationships in the data using charts like histograms and scatterplots.

Most of the tests and techniques we've used so far come from statistics, such as linear regression and the independent samples t-test. What distinguishes data analytics from statistics is not necessarily the *means*, but the *ends*.

Data Analytics

With data analytics, we are less concerned about the methods of analyzing data, and more about using the outcomes to meet some external objective. These can be different: for example, you've seen that while some relationships can be statistically significant, they might not be substantively meaningful for the business.

Data analytics is also concerned with the technology needed to implement these insights. For example, we may need to clean datasets, design dashboards, and disseminate these assets quickly and efficiently. While the focus of this book has been on the *statistical* foundations of analytics, there are other *computational* and *technological* foundations to be aware of, which will be discussed later in this chapter.

Business Analytics

In particular, data analytics is used to guide and meet business objectives and assist business stakeholders; analytics professionals often have one foot in the business operations world and another in the information technology one. The term *business analytics* is often used to describe this combination of duties.

An example of a data or business analytics project might be to analyze movie rental data. Based on exploratory data analysis, the analyst may hypothesize that comedies sell particularly well on holiday weekends. Working with product managers or other business stakeholders, they may run small experiments to collect and further test this hypothesis. Elements of this workflow should sound familiar from earlier chapters of this book.

Data Science

Finally, there is data science: another field that has inseparable ties to statistics, but that is focused on unique outcomes.

Data scientists also commonly approach their work with business objectives in mind, but its scope is quite different from data analytics. Going back to our movie rental example, a data scientist might build an algorithmically powered system to recommend movies to individuals based on what customers similar to them rented. Building and deploying such a system requires considerable engineering skills. While it's unfair to say that data scientists don't have real ties to the business, they are often more aligned with engineering or information technology than their data analytics counterparts.

Machine Learning

To summarize this distinction, we can say that while data analytics is concerned with *describing* and *explaining* data relationships, data science is concerned with building *predictive* systems and products, often using machine learning techniques.

Machine learning is the practice of building algorithms that improve with more data without being explicitly programmed to do so. For example, a bank might deploy machine learning to detect whether a customer will default on a loan. As more data is fed in, the algorithm may find patterns and relationships in the data and use them to better predict the likelihood of a default. Machine learning models can offer incredible predictive accuracy and can be used in a variety of scenarios. That said, it's tempting to build a complex machine learning algorithm when a simple one will suffice, and this can lead to difficulty with interpreting and relying on the model.

Machine learning is beyond the scope of this book; for a fantastic overview, check out Aurélien Géron's *Hands-On Machine Learning with Scikit-Learn, Keras, and Tensor-Flow*, 2nd edition (O'Reilly). That book is conducted heavily in Python, so it's best to have completed Part III of this one first.

Distinct, but Not Exclusive

While distinctions among statistics, data analytics, and data science are meaningful, we shouldn't let them create unnecessary borders. In any of these disciplines, the difference between a categorical and continuous dependent variable is meaningful. All use hypothesis testing to frame problems. We have statistics to thank for this common parlance of working with data.

Data analytics and data science roles are often intermingled as well. In fact, you've learned the basics of a core data science technique in this book: linear regression. In short, there is more that unites these fields than divides them. Though this book is focused on data analytics, you are prepared to explore them all; this will be especially so once you've learned R and Python.

Now that we've contextualized data analytics with statistics and data science, let's do the same for Excel, R, Python, and other tools you may learn in analytics.

The Importance of the Data Analytics Stack

Before technical know-how of any single tool, an analytics professional should have the ability to choose and pair *different* tools given the strengths and weaknesses of each.

It's common for web developers or database administrators to refer to their "stack" of tools used to do the job. We can use the same idea to helpful effect in data analytics. When one tool or "slice" of the stack comes up short, the focus ought not to be on blaming it for shortcomings, but on choosing a different slice or slices. That is, we ought to think of these different slices as complements rather than substitutes.

Figure 5-1 is my conceptualization of the four slices of the analytics stack. This is a vast oversimplification of what data tools are used in organizations; mapping out end-to-end analytics pipelines can get complicated. The slices are arranged in order from where data is stored and maintained by information technology departments (databases) to where it is used and explored by business end users (spreadsheets). Any of these slices can be used together to craft solutions.

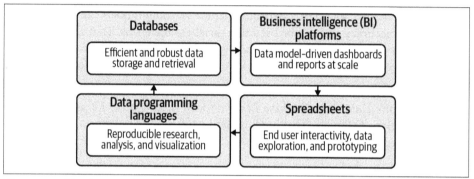

Figure 5-1. The data analytics stack

Let's take some time to explore each slice of the stack. I'll cover these slices from what I assume is the most familiar to the least familiar for the typical reader.

Spreadsheets

I won't spend too much time on what spreadsheets are and how they work; you're pretty accustomed to them by now. These principles apply to other spreadsheet applications like Google Sheets, LibreOffice, and more; we've been focused on Excel in this book, so I'll emphasize it here. You've seen that the spreadsheet can bring analytics to life and is a great tool for EDA. This ease of use and flexibility makes spreadsheets ideal for distributing data to end users.

But this flexibility can be both a virtue and a vice. Have you ever built a spreadsheet model where you landed at some number, only to reopen the file a few hours later and inexplicably get a different number? Sometimes it can feel like playing whack-a-mole with spreadsheets; it's very hard to isolate one layer of the analysis without affecting the others.

A well-designed data product looks something like what's shown in Figure 5-2:

- The raw data is distinct and untouched by the analysis.
- The data is then processed for any relevant cleanup and analysis.
- Any resulting charts or tables are isolated output.

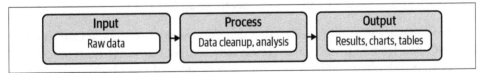

Figure 5-2. Input, process, output

Although there are principles to follow with this approach in spreadsheets, these layers tend to become a messy soup: users may write directly over raw data, or build calculations atop calculations to the point where it becomes very difficult to track all the references pointing to a given cell. Even with a solid workbook design in place, it can still be hard to achieve the ultimate goal of an input-process-output model, which is *reproducibility*. The idea here is that, given the same input and processes, the same output will be achieved time and again. Clearly, a workbook isn't reproducible when, due to error-prone steps, clunky calculations, or more, it's not guaranteed you'll arrive at the same outcome each time you open the file.

Messy or nonreproducible workbooks have spawned horror stories in every field from food service to finance regulation: for a frightening overview, check out this article from the European Spreadsheet Risks Interest Group (*https://oreil.ly/gWWw3*). Maybe the analysis you do isn't as high-stakes as trading bonds or publishing groundbreaking academic research. But nobody likes slow, error-prone processes that produce unreliable results. But enough doom-mongering; as I hope to continuously emphasize, Excel and other spreadsheets have their rightful place in analytics. Let's take a look at some tools that help build clean, reproducible workflows in Excel.

VBA

You will see that in general, reproducibility is achieved in computing by recording each step of the analysis as code, which can be saved and quickly re-executed later. Excel does indeed have an in-house programming language in Visual Basic for Applications (VBA).

Although VBA does allow for a process to be recorded as code, it lacks many of the features of a full-on statistical programming language: in particular, the abundance of free packages for specialized analysis. Moreover, Microsoft has all but sunset VBA, moving resources to its new Office Scripts language as a built-in Excel automation tool, as well as JavaScript and, if rumors are to be believed, Python.

Modern Excel

I'll use this term for the series of tools centered on business intelligence (BI) that Microsoft began releasing to Excel in Excel 2010. These tools are incredibly powerful and fun to use, and they bust many of the myths about what Excel can and can't do. Let's take a look at the three applications that make up Modern Excel:

- Power Query is a tool for extracting data from various sources, transforming it, and then loading it into Excel. These data sources can range from *.csv* files to relational databases and can contain many millions of records: while it still may be true that an Excel workbook itself can only contain about a million rows, it can contain several times that limit if read via Power Query.

  What's even better, Power Query is *fully reproducible* on account of Microsoft's M programming language. Users can add and edit steps via a menu, which generates the M code, or write it themselves. Power Query is a showstopping force for Excel; not only does it blow away previous constraints on how much data can flow through a workbook, it makes the retrieval and manipulation of this data fully reproducible.

- Power Pivot is a relational data modeling tool for Excel. We'll discuss relational data models in more depth later in this chapter when we review databases.

- Finally, Power View is a tool for creating interactive charts and visualizations in Excel. This is especially helpful for building dashboards.

I highly suggest you take some time to learn about Modern Excel, particularly if you are in a role that relies highly on it for analysis and reporting. Many of the naysaying claims about Excel, such as that it can't handle more than a million rows, or work with diverse data sources, are no longer true with these releases.

That said, these tools aren't necessarily built to conduct statistical analysis, but to aid in other analytics roles, such as building reports and disseminating data. Fortunately, there is ample room to mix Power Query and Power Pivot with tools like R and Python to build exceptional data products.

Despite this evolution and its many benefits, Excel is frowned upon by many in the analytics world because of the misfortunes its overuse can lead to. This leads us to ask: *why is Excel overused in the first place?* It's because business users, for lack of better alternatives and resources, see it as an intuitive, flexible place for storing and analyzing data.

I say, if you can't beat 'em, join 'em: Excel *is* a great tool for exploring and interacting with data. It's even, with its latest features, become a great tool for building reproducible data-cleaning workflows and relational data models. But there are some analytics functions that Excel is not so great for, such as storing mission-critical data, distributing dashboards and reports across multiple platforms, and performing advanced statistical analysis. For those, let's look at the alternatives.

Databases

Databases, specifically *relational* databases, are a relatively ancient technology in the world of analytics, with their origins tracing to the early 1970s. The building block of relational databases is something you've seen before: the *table*. Figure 5-3 is such an example: we've been referring to columns and rows of such a table with the statistical terms of *variables* and *observations*. Their counterparts in the language of databases are *fields* and *records*.

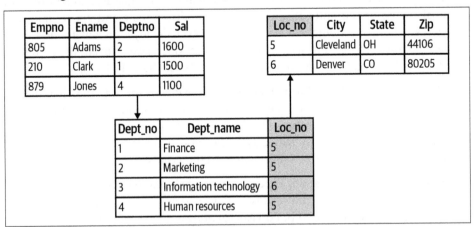

Dept_no	Dept_name	Loc_no
1	Finance	5
2	Marketing	5
3	Information technology	6
4	Human resources	5

Column, variable, *field*

Row, observation, *record*

Figure 5-3. A labeled database table

If you were asked to connect data from Figure 5-4 together, you may use Excel's VLOOKUP() function, using shared columns as "lookup fields" to transfer data from one table to another. There's a lot more to it, but this is the crux of *relational data models*: to use relations between data across tables to store and manage data efficiently. I like to call VLOOKUP() the *duct tape of Excel* because of its ability to connect datasets together. If VLOOKUP() is duct tape, then relational data models are welders.

Empno	Ename	Deptno	Sal
805	Adams	2	1600
210	Clark	1	1500
879	Jones	4	1100

Loc_no	City	State	Zip
5	Cleveland	OH	44106
6	Denver	CO	80205

Dept_no	Dept_name	Loc_no
1	Finance	5
2	Marketing	5
3	Information technology	6
4	Human resources	5

Figure 5-4. Relationships among fields and tables in a relational database

A *relational database management system* (RDBMS) is designed to exploit this basic concept for large-scale data storage and retrieval. When you place an order at a store or sign up for a mailing list, that data likely passes through an RDBMS. While built on the same concepts, Power Pivot's use case is more for BI analysis and reporting and is not a full-service RDMBS.

Structured Query Language, or SQL, is traditionally used to interact with databases. This is another crucial topic in analytics that is outside our book's scope; for a great introduction, check out Alan Beaulieu's *Learning SQL*, 3rd edition (O'Reilly). Keep in mind that while "SQL" (or "sequel") as a language name is often used generically, several "dialects" exist depending on the RDBMS of interest. Some of these systems, like Microsoft or Oracle, are proprietary; others, like PostgreSQL or SQLite, are open source.

A classic acronym for the operations SQL can perform is *CRUD*, or Create, Read, Update, Delete. As a data analyst, you'll most typically be involved with *reading* data from a database rather than changing it. For these operations, the difference in SQL dialects will be negligible across various platforms.

Business Intelligence Platforms

This is an admittedly broad swath of tools and likely the most ambiguous slice of the stack. Here I mean enterprise tools that allow users to gather, model, and display data. Data warehousing tools like MicroStrategy and SAP BusinessObjects straddle the line here, since they are tools designed for self-service data gathering and analysis. But these often have limited visualization and interactive dashboard-building included.

That's where tools like Power BI, Tableau, and Looker come in. These platforms, nearly all proprietary, allow users to build data models, dashboards, and reports with minimal coding. Importantly, they make it easy to disseminate and update information across an organization; these assets are often even deployed to tablets and smartphones in a variety of formats. Many organizations have moved their routine reporting and dashboard creation from spreadsheets into these BI tools.

For all their benefits, BI platforms tend to be inflexible in the way they handle and visualize data. With the objective to be straightforward for business users and difficult to break, they often lack the features that seasoned data analysts need to do the necessary "hacking" for the task at hand. They can also be expensive, with single-user annual licenses running several hundred or even thousands of dollars.

Given what you've learned about Excel, it's worth pointing out that the elements of Modern Excel (Power Query, Power Pivot, Power View) are also available for Power BI. What's more, it's possible to build visualizations in Power BI using R and Python code. Other BI systems have similar capabilities; I've been focusing on Power BI here due to our earlier focus on Excel.

Data Programming Languages

That brings us to our final slice: data programming languages. By this I mean wholly scripted software applications used specifically for data analytics. Many analytics professionals do phenomenal work without this slice in their stack. Moreover, many vendor tools are moving toward low- or no-code solutions for sophisticated analytics.

All that said, I strongly encourage you to learn how to code. It will sharpen your understanding of how data processing works, and give you fuller control of your workflow than relying on a graphical user interface (GUI), or point-and-click software.

For data analytics, two open source programming languages are good fits: R and Python, hence the focus of the rest of this book. Each includes a dizzying universe of free packages made to help with everything from social media automation to geospatial analysis. Learning these languages opens the door to advanced analytics and data science. If you thought Excel was a powerful way to explore and analyze data, wait until you get the hang of R and Python.

On top of that, these tools are ideal for reproducible research. Think back to Figure 5-2 and the difficulties we spotted in separating these steps in Excel. As programming languages, R and Python record all steps taken in an analysis. This workflow leaves the raw data intact by first reading from external sources, *then* operating on a copy of that data. The workflow also makes it easier to track changes and contributions to files by a process known as *version control*, which will be discussed in Chapter 14.

R and Python are open source software applications, which means that their source code is freely available for anyone to build on, distribute, or contribute to. This is quite different from Excel, which is a proprietary offering. Both open source and proprietary systems have their advantages and disadvantages. In the case of R and Python, allowing anyone to develop freely on the source code has led to a rich ecosystem of packages and applications. It's also lowered the barrier to entry for newcomers to get involved.

All that said, it's common to find that critical parts of open source infrastructure are maintained by developers in their spare time without any compensation. It may not be ideal to rely on continued development and maintenance of an infrastructure that is not commercially guaranteed. There are ways to mitigate this risk; in fact, many companies exist solely to support, maintain, and augment open source systems. You will see this relationship at work in our later discussions on R and Python; it may surprise you that it's entirely possible to make money by providing services based on freely available code.

The "data programming language" slice of the stack likely has the steepest learning curve of them all: after all, it's literally learning a new language. Learning *one* such language may sound like a stretch, so how and why on earth will you be learning two?

First of all, as we mentioned at the beginning of the book, you're not starting at zero. You have strong knowledge of how to program and how to work with data. So take it in stride that you *have* learned how to code...sort of.

 There's a benefit to being multilingual with data programming languages, just as there is with spoken languages. At a pragmatic level, employers may use either of them, so it's smart to cover your bases. But you're not just ticking a box by learning both: each language has its own unique features, and you may find it easier to use one for a given use case. Just as it's smart to think about different slices of the stack as complements and not subsitutes, the same attitude holds for tools in the same slice.

Conclusion

Data analysts often wonder which tools they should focus on learning or becoming the expert in. I would suggest not becoming the expert in any single tool, but in learning different tools from each slice of the stack well enough to contextualize and choose between them. Seen from this perspective, it makes little sense to claim one slice of the stack as inferior to another. They are meant to be complementary, not substitutes.

In fact, many of the most powerful analytics products come from *combining* slices of the stack. For example, you might use Python to automate the production of Excel-based reports, or pull data from an RDBMS into a BI platform's dashboard. Although these use cases are beyond the scope of this book, the upshot for our discussion is: *don't ignore Excel*. It's a valued slice of the stack that is only complemented by your skills in R and Python.

In this book, we focus on spreadsheets (Excel) and data programming languages (R and Python). These tools are particularly suited for the statistically based roles of data analytics, which as we've discussed have some overlap with traditional statistics and with data science. But as we've also discussed, analytics involves more than pure statistical analysis, and relational databases and BI tools can be helpful for these duties. Once you've familiarized yourself with the topics in this book, consider rounding out your knowledge of the data analytics stack with the titles I suggested earlier in this chapter.

What's Next

With this big-picture overview of data analytics and data analytics applications in mind, let's dive into exploring new tools.

We'll start with R because I consider it a more natural jumping-off point into data programming for Excel users. You'll learn how to conduct much of the same EDA and hypothesis testing with R that you did with Excel, which will put you in a great position for more advanced analytics. Then you'll do the same with Python. At each point along the way, I'll help relate what you're learning to what you already know, so that you see how familiar so many of the concepts actually are. See you in Chapter 6.

Exercises

This chapter is more conceptual than applied, so there are no exercises. I encourage you to come back to it as you branch out into other areas of analytics and relate them to each other. When you encounter a new data tool at work or while perusing social media or industry publications, ask yourself which slices of the stack it covers, whether it's open source, and so forth.

## PART II

From Excel to R

## CHAPTER 6

First Steps with R for Excel Users

In Chapter 1 you learned how to conduct exploratory data analysis in Excel. You may recall from that chapter that John Tukey is credited with popularizing the practice of EDA. Tukey's approach to data inspired the development of several statistical programming languages, including S at the legendary Bell Laboratories. In turn, S inspired R. Developed in the early 1990s by Ross Ihaka and Robert Gentleman, the name is a play both on its derivation from S and its cofounders' first names. R is open source and maintained by the R Foundation for Statistical Computing. Because it was built primarily for statistical computation and graphics, it's most popular among researchers, statisticians, and data scientists.

R was developed specifically with statistical analysis in mind.

Downloading R

To get started, navigate to the R Project's website (*https://r-project.org*). Click the link at the top of the page to download R. You will be asked to choose a mirror from the Comprehensive R Archive Network (CRAN). This is a network of servers that distributes R source code, packages, and documentation. Choose a mirror near you to download R for your operating system.

Getting Started with RStudio

You've now installed R, but we will make one more download to optimize our coding experience. In Chapter 5, you learned that when software is open source, anyone is free to build on, distribute, or contribute to it. For example, vendors are welcome to offer an *integrated development environment* (IDE) to interact with the code. The RStudio IDE combines tools for code editing, graphics, documentation, and more under a single interface. This has become the predominant IDE for R programming in its decade or so on the market, with users building everything from interactive dashboards (Shiny) to research reports (R Markdown) with its suite of products.

You may be wondering, *if RStudio is so great, why did we bother installing R?* These are in fact two distinct downloads: we downloaded R for the *code base*, and RStudio for an *IDE to work with the code*. This decoupling of applications may be unfamiliar to you as an Excel user, but it's quite common in the open source software world.

 RStudio is a platform to *work with* R code, not the code base itself. First, download R from CRAN; then download RStudio.

To download RStudio, head to the download page (*https://oreil.ly/rfP1X*) of its website. You will see that RStudio is offered on a tiered pricing system; select the free RStudio Desktop. (RStudio is an excellent example of how to build a solid business on top of open source software.) You'll come to love RStudio, but it can be quite overwhelming at first with its many panes and features. To overcome this initial discomfort, we'll take a guided tour.

First, head to the home menu and select File → New File → R Script. You should now see something like Figure 6-1. There are lots of bells and whistles here; the idea of an IDE is to have all the tools needed for code development in one place. We'll cover the features in each of the four panes that you should know to get started.

Located in the lower left-hand corner of RStudio, the *console* is where commands are submitted to R to execute. Here you will see the > sign followed by a blinking cursor. You can type operations here and then press Enter to execute. Let's start with something very basic, like finding 1 + 1, as in Figure 6-2.

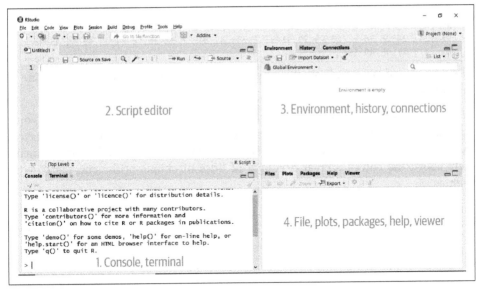

Figure 6-1. The RStudio IDE

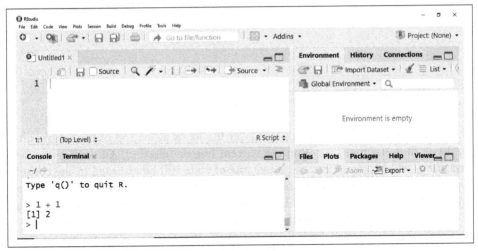

Figure 6-2. Coding in RStudio, starting with 1 + 1

You may have noticed that a [1] appears before your result of 2. To understand what this means, type and execute 1:50 in the console. The : operator in R will produce all numbers in increments of 1 between a given range, akin to the fill handle in Excel. You should see something like this:

```
1:50
#> [1]  1  2  3  4  5  6  7  8  9 10 11 12 13 14 15 16 17 18 19 20 21 22 23
#> [24] 24 25 26 27 28 29 30 31 32 33 34 35 36 37 38 39 40 41 42 43 44 45 46
#> [47] 47 48 49 50
```

These bracketed labels indicate the numeric position of the first value for each line in the output.

While you can continue to work from here, it's often a good idea to first write your commands in a *script*, and then send them to the console. This way you can save a long-term record of the code you ran. The script editor is found in the pane immediately above the console. Enter a couple of lines of simple arithmetic there, as in Figure 6-3.

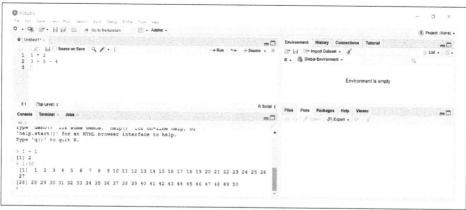

Figure 6-3. Working with the script editor in RStudio

Place your cursor in line 1, then hover over the icons at the top of the script editor until you find one that says "Run the current line or selection." Click that icon and two things will happen. First, the active line of code will be executed in the console. The cursor will also drop to the next line in the script editor. You can send multiple lines to the console at once by selecting them and clicking that icon. The keyboard shortcut for this operation is Ctrl + Enter for Windows, Cmd + Return for Mac. As an Excel user, you're probably a keyboard shortcut enthusiast; RStudio has an abundance of them, which can be viewed by selecting Tools → Keyboard Shortcuts Help.

Let's save our script. From the menu head to File → Save. Name the file *ch-6*. The file extension for R scripts is *.r*. The process of opening, saving, and closing R scripts may remind you of working with documents in a word processor; after all, they are both written records.

We'll now head to the lower-right pane. You will see five tabs here: Files, Plots, Packages, Help, Viewer. R provides plenty of help documentation, which can be viewed in this pane. For example, we can learn more about an R function with the ? operator.

As an Excel user, you know all about functions such as VLOOKUP() or SUMIF(). Some R functions are quite similar to those of Excel; let's learn, for example, about R's square-root function, sqrt(). Enter the following code into a new line of your script and run it using either the menu icon or the keyboard shortcut:

```
?sqrt
```

A document titled "Miscellaneous Mathematical Functions" will appear in your Help window. This contains important information about the sqrt() function, the arguments it takes, and more. It also includes this example of the function in action:

```
require(stats) # for spline
require(graphics)
xx <- -9:9
plot(xx, sqrt(abs(xx)),  col = "red")
lines(spline(xx, sqrt(abs(xx)), n=101), col = "pink")
```

Don't worry about making sense of this code right now; just copy and paste the selection into your script, highlighting the complete selection, and run it. A plot will now appear as in Figure 6-4. I've resized my RStudio panes to make the plot larger. You will learn how to build R plots in Chapter 8.

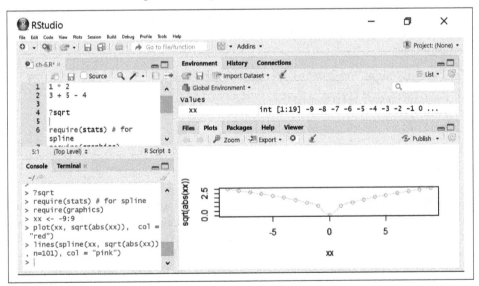

Figure 6-4. Your first R plot

Now, look to your upper-right pane: Environment, History, Connections. The Environment tab lists something called xx next to what looks to be some set of integers. What is this? As it turns out, *you* created this with the code I told you to run blindly from the sqrt() documentation. In fact, much of what we do in R will focus around what is shown here: an *object*.

As you likely noticed, there are several panes, icons, and menu options we overlooked in this tour of RStudio. It's such a feature-rich IDE: don't be afraid to explore, experiment, and search-engine your way to learning more. But for now, you know enough about getting around in RStudio to begin learning R programming proper. You've already seen that R can be used as a fancy calculator. Table 6-1 lists some common arithmetic operators in R.

Table 6-1. Common arithmetic operators in R

Operator	Description
+	Addition
-	Subtraction
*	Multiplication
/	Division
^	Exponent
%%	Modulo
%/%	Floor division

You may be less familiar with the last two operators in Table 6-1: the *modulo* returns the remainder of a division, and *floor division* rounds the division's result down to the nearest integer.

Like Excel, R follows the order of operations in arithmetic.

```
# Multiplication before addition
3 * 5 + 6
#> [1] 21

# Division before subtraction
2 / 2 - 7
#> [1] -6
```

What's the deal with the lines containing the hash (#) and text? Those are *cell comments* used to provide verbal instructions and reminders about the code. Comments help other users—and ourselves at a later date—understand and remember what the code is used for. R does not execute cell comments: this part of the script is for the programmer, not the computer. Though comments can be placed to the right of code, it's preferred to place them above:

```
1 * 2 # This comment is possible
#> [1] 2

# This comment is preferred
2 * 1
#> [1] 2
```

You don't need to use comments to explain *everything* about what your code is doing, but do explain your reasoning and assumptions. Think of it as, well, a commentary. I will continue to use comments in this book's examples where relevant and helpful.

 Get into the habit of including comments to document your objectives, assumptions, and reasoning for writing the code.

As previously mentioned, functions are a large part of working in R, just as in Excel, and often look quite similar. For example, we can take the absolute value of −100:

```
# What is the absolute value of -100?
abs(-100)
#> [1] 100
```

However, there are some quite important differences for working with functions in R, as these errors indicate.

```
# These aren't going to work
ABS(-100)
#> Error in ABS(-100) : could not find function "ABS"
Abs(-100)
#> Error in Abs(-100) : could not find function "Abs"
```

In Excel, you can enter the ABS() function as lowercase abs() or proper case Abs() without a problem. In R, however, the abs() function *must* be lowercase. This is because R is *case-sensitive*. This is a major difference between Excel and R, and one that is sure to trip you up sooner or later.

 R is a case-sensitive language: the SQRT() function is not the same as sqrt().

Like in Excel, some R functions, like sqrt(), are meant to work with numbers; others, like toupper(), work with characters:

```
# Convert to upper case
toupper('I love R')
#> [1] "I LOVE R"
```

Let's look at another case where R behaves similarly to Excel, with one exception that will have huge implications: comparison operators. This is when we compare some relationship between two values, such as whether one is greater than the other.

```
# Is 3 greater than 4?
3 > 4
#> [1] FALSE
```

R will return a TRUE or FALSE as a result of any comparison operator, just as would Excel. Table 6-2 lists R's comparison operators.

Table 6-2. Comparison operators in R

Operator	Meaning
>	Greater than
<	Less than
>=	Greater than or equal to
<=	Less than or equal to
!=	Not equal to
==	Equal to

Most of these probably look familiar to you, except...did you catch that last one? That's correct, you do not check whether two values are equal to one another in R with one equal sign, but with *two*. This is because a single equal sign in R is used to *assign objects*.

Objects Versus Variables

Stored objects are also sometimes referred to as *variables* because of their ability to be overwritten and change values. However, we've already been referring to *variables* in the statistical sense throughout this book. To avoid this confusing terminology, we will continue to refer to "objects" in the programming sense and "variables" in the statistical.

If you're not quite sure what the big deal is yet, bear with me for another example. Let's assign the absolute value of –100 to an object; we'll call it my_first_object.

```
# Assigning an object in R
my_first_object = abs(-100)
```

You can think of an object as a shoebox that we are putting a piece of information into. By using the = operator, we've stored the result of abs(-100) in a shoebox called my_first_object. We can open this shoebox by *printing* it. In R, you can simply do this by running the object's name:

```
# Printing an object in R
my_first_object
#> [1] 100
```

Another way to assign objects in R is with the <- operator. In fact, this is usually preferred to = in part to avoid the confusion between it and ==. Try assigning another object using this operator, then printing it. The keyboard shortcut is Alt+- (Alt +minus) on Windows, and Option-- (Option-minus) on Mac. You can get creative with your functions and operations, like I did:

```
my_second_object <- sqrt(abs(-5 ^ 2))
my_second_object
#> [1] 5
```

Object names in R must start with a letter or dot and should contain only letters, numbers, underscores, and periods. There are also a few off-limit keywords. That leaves a lot of margin for "creative" object naming. But good object names are indicative of the data they store, similar to how the label on a shoebox signals what kind of shoe is inside.

R and Programming Style Guides

Some individuals and organizations have consolidated programming conventions into "style guides," just as a newspaper might have a style guide for writing. These style guides cover which assignment operators to use, how to name objects, and more. One such R style guide has been developed by Google and is available online (*https://oreil.ly/fAeJi*).

Objects can contain different types or *modes* of data, just as you might have different categories of shoeboxes. Table 6-3 lists some common data types.

Table 6-3. Common data types in R

Data type	Example
Character	'R', 'Mount', 'Hello, world'
Numeric	6.2, 4.13, 3
Integer	3L, -1L, 12L
Logical	TRUE, FALSE, T, F

Let's create some objects of different modes. First, character data is often enclosed in single quotations for legibility, but double quotes also work and can be particularly helpful if you want to include a single quote as part of the input.

```
my_char <- 'Hello, world'
my_other_char <- "We're able to code R!"
```

Numbers can be represented as decimals or whole numbers:

```
my_num <- 3
my_other_num <- 3.21
```

However, whole numbers can also be stored as a distinct integer data type. The L included in the input stands for *literal*; this term comes from computer science and is used to refer to notations for fixed values:

```
my_int <- 12L
```

T and F will by default evaluate as logical data to TRUE and FALSE, respectively:

```
my_logical <- FALSE
my_other_logical <- F
```

We can use the str() function to learn about the *structure* of an object, such as its type and the information contained inside:

```
str(my_char)
#> chr "Hello, world"
str(my_num)
#> num 3
str(my_int)
#> int 12
str(my_logical)
#> logi FALSE
```

Once assigned, we are free to use these objects in additional operations:

```
# Is my_num equal to 5.5?
my_num == 5.5
#> [1] FALSE

# Number of characters in my_char
nchar(my_char)
#> [1] 12
```

We can even use objects as input in assigning other objects, or reassign them:

```
my_other_num <- 2.2
my_num <- my_num/my_other_num
my_num
#> [1] 1.363636
```

"So what?" you may be asking. "I work with a lot of data, so how is assigning each number to its own object going to help me?" Fortunately, you'll see in Chapter 7 that it's possible to combine multiple values into one object, much as you might do with ranges and worksheets in Excel. But before that, let's change gears for a moment to learn about packages.

Packages in R

Imagine if you weren't able to download applications on your smartphone. You could make phone calls, browse the internet, and jot notes to yourself—still pretty handy. But the real power of a smartphone comes from its applications, or apps. R ships much like a "factory-default" smartphone: it's still quite useful, and you could accomplish nearly anything necessary with it if you were forced to. But it's often more efficient to do the R equivalent of installing an app: *installing a package*.

The factory-default version of R is called "base R." Packages, the "apps" of R, are shareable units of code that include functions, datasets, documentation, and more. These packages are built on top of base R to improve functionality and add new features.

Earlier, you downloaded base R from CRAN. This network also hosts over 10,000 packages that have been contributed by R's vast user base and vetted by CRAN volunteers. This is your "app store" for R, and to repurpose the famous slogan, "There's a package for that." While it's possible to download packages elsewhere, it's best as a beginner to stick with what's hosted on CRAN. To install a package from CRAN, you can run `install.packages()`.

CRAN Task Views

It can be difficult for a newcomer to identify the right R packages for their needs. Fortunately, the CRAN team provides something like "curated playlists" of packages for given use cases with CRAN Task Views (*https://oreil.ly/q31wg*). These are bundles of packages meant to assist with everything from econometrics to genetics and provide a great landscape of helpful R packages. As you continue to learn the language, you'll get more comfortable locating and sizing up the right package for your requirements.

We'll be using packages in this book to help us with tasks like data manipulation and visualization. In particular, we'll be using the `tidyverse`, which is actually a *collection* of packages designed to be used together. To install this collection, run the following in the console:

```
install.packages('tidyverse')
```

You've just installed a number of helpful packages; one of which, `dplyr` (usually pronounced *d-plier*), includes a function `arrange()`. Try opening the documentation for this function and you'll receive an error:

```
?arrange
#> No documentation for 'arrange' in specified packages and libraries:
#> you could try '??arrange'
```

To understand why R can't find this `tidyverse` function, go back to the smartphone analogy: even though you've installed an app, you still need to open it to use it. Same with R: we've installed the package with `install.packages()`, but now we need to call it into our session with `library()`:

```
# Call the tidyverse into our session
library(tidyverse)
#> -- Attaching packages ------------------------ tidyverse  1.3.0 --
#> v ggplot2 3.3.2     v purrr   0.3.4
#> v tibble  3.0.3     v dplyr   1.0.2
#> v tidyr   1.1.2     v stringr 1.4.0
#> v readr   1.3.1     v forcats 0.5.0
#> -- Conflicts ------------------------- tidyverse_conflicts() --
#> x dplyr::filter() masks stats::filter()
#> x dplyr::lag()    masks stats::lag()
```

The packages of `tidyverse` are now available for the rest of your R session; you can now run the example without error.

Packages are *installed* once, but *called* for each session.

Upgrading R, RStudio, and R Packages

RStudio, R packages, and R itself are constantly improving, so it's a good idea to occasionally check for updates. To update RStudio, navigate to the menu and select Help → Check for Updates. If you're due for an update, RStudio will guide you through the steps.

To update all packages from CRAN, you can run this function and follow the prompted steps:

```
update.packages()
```

You can also update packages from the RStudio menu by going to Tools → Check for Package Updates. An Update Packages menu will appear; select all of the packages that you wish to update. You can also install packages via the Tools menu.

Upgrading R itself is unfortunately more involved. If you are on a Windows computer, you can use the `updateR()` function from the package `installr` and follow its instructions:

```
# Update R for Windows
install.packages('installr')
library(installr)
updateR()
```

For Mac, return to the CRAN website (*https://cran.r-project.org*) to install the latest version of R.

Conclusion

In this chapter, you learned how to work with objects and packages in R and got the hang of working with RStudio. You've learned a lot; I think it's time for a break. Go ahead and save your R script and close out of RStudio by selecting File → Quit Session. When you do so, you'll be asked: "Save workspace image to ~/.RData?" As a rule, *don't save your workspace image*. If you do, a copy of all saved objects will be saved so that they'll be available for your next session. While this *sounds* like a good idea, it can get cumbersome to store these objects and keep track of *why* you stored them in the first place.

Instead, rely on the R script itself to regenerate these objects in your next session. After all, the advantage of a programming language is that it's reproducible: no need to drag objects around with us if we can create them on demand.

 Err on the side of *not* saving your workspace image; you should be able to re-create any objects from a previous session using your script.

To prevent RStudio from preserving your workspace between sessions, head to the home menu and go to Tools → Global Options. Under the General menu, change the two settings under Workspace as shown in Figure 6-5.

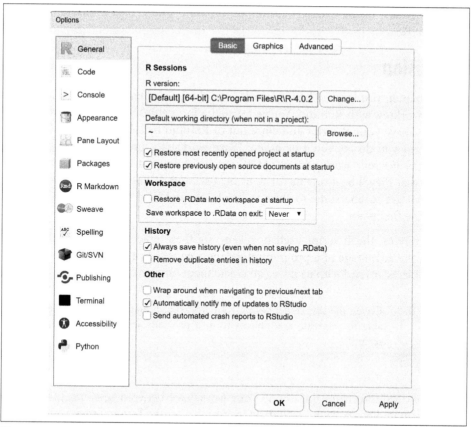

Figure 6-5. Customized workspace options in RStudio

Exercises

The following exercises provide additional practice and insight on working with objects, packages, and RStudio:

1. In addition to being a workhorse of a tool, RStudio provides endless appearance customizations. From the menu, select Tools → Global Options → Appearance and customize your editor's font and theme. For example, you may decide to use a "dark mode" theme.

2. Using a script in RStudio, do the following:
   - Assign the sum of 1 and 4 as a.
   - Assign the square root of a as b.
   - Assign b minus 1 as d.
   - What type of data is stored in d?
   - Is d greater than 2?

3. Install the psych package from CRAN, and load it into your session. Use comments to explain the differences between installing and loading a package.

Along with these exercises, I encourage you to begin using R immediately in your day-to-day work. For now, this may just involve using the application as a fancy calculator. But even this will help you get comfortable with using R and RStudio.

## CHAPTER 7

Data Structures in R

Toward the end of Chapter 6 you learned how to work with packages in R. It's common to load any necessary packages at the beginning of a script so that there are no surprises about required downloads later on. In that spirit, we'll call in any packages needed for this chapter now. You may need to install some of these; if you need a refresher on doing that, look back to Chapter 6. I'll further explain these packages as we get to them.

```
# For importing and exploring data
library(tidyverse)

# For reading in Excel files
library(readxl)

# For descriptive statistics
library(psych)

# For writing data to Excel
library(writexl)
```

Vectors

In Chapter 6 you also learned about calling functions on data of different modes, and assigning data to objects:

```
my_number <- 8.2
sqrt(my_number)
#> [1] 2.863564

my_char <- 'Hello, world'
toupper(my_char)
#> [1] "HELLO, WORLD"
```

Chances are, you generally work with more than one piece of data at a time, so assigning each to its own object probably doesn't sound too useful. In Excel, you can place data into contiguous cells, called a *range*, and easily operate on that data. Figure 7-1 depicts some simple examples of operating on ranges of both numbers and text in Excel:

	A	B	C	D	E	F	G	H
1	Billy	BILLY	=UPPER(A1)		5	8	2	7
2	Jack	JACK	=UPPER(A2)		2.236067977	2.828427125	1.414213562	2.645751311
3	Jill	JILL	=UPPER(A3)		=SQRT(E1)	=SQRT(F1)	=SQRT(G1)	=SQRT(H1)
4	Johnny	JOHNNY	=UPPER(A4)					
5	Susie	SUSIE	=UPPER(A5)					
6								

Figure 7-1. Operating on ranges in Excel

Earlier I likened the *mode* of an object to a particular type of shoe in a shoebox. The *structure* of an object is the shape, size, and architecture of the shoebox itself. In fact, you've already been finding the structure of an R object with the str() function.

R contains several object structures: we can store and operate on a bit of data by placing it in a particular structure called a *vector*. Vectors are collections of one or more elements of data of the same type. Turns out we've already been using vectors, which we can confirm with the is.vector() function:

```
is.vector(my_number)
#> [1] TRUE
```

Though my_number is a vector, it only contains one element—sort of like a single cell in Excel. In R, we would say this vector has a length of 1:

```
length(my_number)
#> [1] 1
```

We can make a vector out of multiple elements, akin to an Excel range, with the c() function. This function is so called because it serves to *combine* multiple elements into a single vector. Let's try it:

```
my_numbers <- c(5, 8, 2, 7)
```

This object is indeed a vector, its data is numeric, and it has a length of 4:

```
is.vector(my_numbers)
#> [1] TRUE

str(my_numbers)
#> [1] num [1:4] 5 8 2 7

length(my_numbers)
#> [1] 4
```

Let's see what happens when we call a function on `my_numbers`:

```
sqrt(my_numbers)
#> [1] 2.236068 2.828427 1.414214 2.645751
```

Now we're getting somewhere. We could similarly operate on a character vector:

```
roster_names <- c('Jack', 'Jill', 'Billy', 'Susie', 'Johnny')
toupper(roster_names)
#> [1] "JACK"   "JILL"   "BILLY"  "SUSIE"  "JOHNNY"
```

By combining elements of data into vectors with the `c()` function, we were able to easily reproduce in R what was shown in Excel in Figure 7-1. What happens if elements of different types are assigned to the same vector? Let's give it a try:

```
my_vec <- c('A', 2, 'C')
my_vec
#> [1] "A" "2" "C"

str(my_vec)
#> chr [1:3] "A" "2" "C"
```

R will *coerce* all elements to be of the same type so that they can be combined into a vector; for example, the numeric element 2 in the previous example was coerced into a character.

Indexing and Subsetting Vectors

In Excel, the `INDEX()` function serves to find the position of an element in a range. For example, I will use `INDEX()` in Figure 7-2 to extract the element in the third position of the named range `roster_names` (cells `A1:A5`):

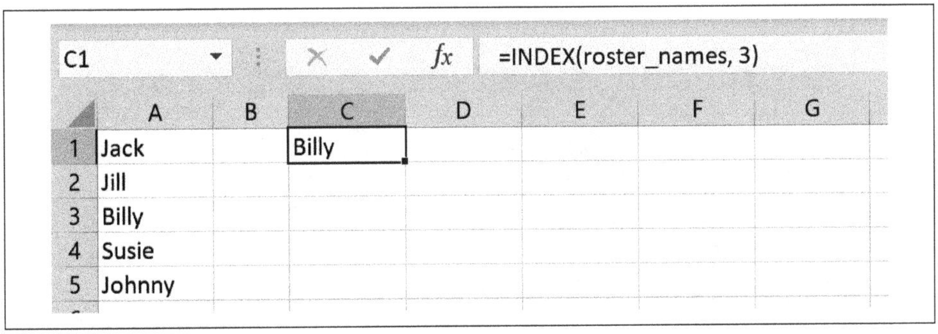

Figure 7-2. The `INDEX()` function on an Excel range

We can similarly index a vector in R by affixing the desired index position inside brackets to the object name:

```
# Get third element of roster_names vector
roster_names[3]
#> [1] "Billy"
```

Using this same notation, it's possible to select multiple elements by their index number, which we'll call *subsetting*. Let's again use the : operator to pull all elements between position 1 and 3:

```
# Get first through third elements
roster_names[1:3]
#> [1] "Jack"  "Jill"  "Billy"
```

It's possible to use functions here, too. Remember length()? We can use it to get everything through the last element in a vector:

```
# Get second through last elements
roster_names[2:length(roster_names)]
#> [1] "Jill"   "Billy"  "Susie"  "Johnny"
```

We could even use the c() function to index by a vector of nonconsecutive elements:

```
# Get second and fifth elements
roster_names[c(2, 5)]
#> [1] "Jill"    "Johnny"
```

From Excel Tables to R Data Frames

"This is all well and good," you may be thinking, "but I don't just work with small ranges like these. What about whole data *tables*?" After all, in Chapter 1 you learned all about the importance of arranging data into variables and observations, such as the *star* data shown in Figure 7-3. This is an example of a *two-dimensional* data structure.

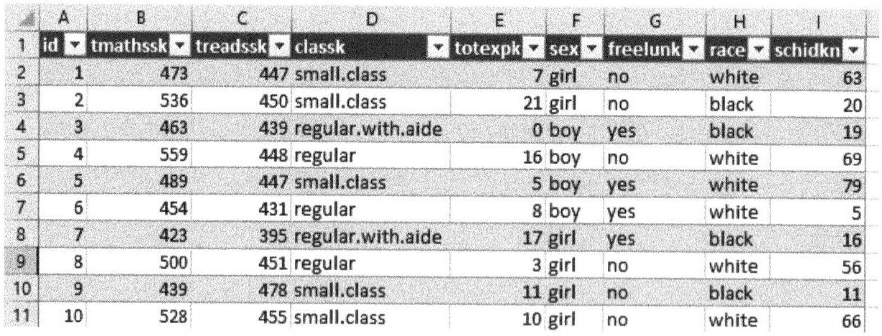

	A	B	C	D	E	F	G	H	I
1	id	tmathssk	treadssk	classk	totexpk	sex	freelunk	race	schidkn
2	1	473	447	small.class	7	girl	no	white	63
3	2	536	450	small.class	21	girl	no	black	20
4	3	463	439	regular.with.aide	0	boy	yes	black	19
5	4	559	448	regular	16	boy	no	white	69
6	5	489	447	small.class	5	boy	yes	white	79
7	6	454	431	regular	8	boy	yes	white	5
8	7	423	395	regular.with.aide	17	girl	yes	black	16
9	8	500	451	regular	3	girl	no	white	56
10	9	439	478	small.class	11	girl	no	black	11
11	10	528	455	small.class	10	girl	no	white	66

Figure 7-3. A two-dimensional data structure in Excel

Whereas R's vector is one-dimensional, the *data frame* allows for storing data in both rows *and* columns. This makes the data frame the R equivalent of an Excel table. Put formally, a data frame is a two-dimensional data structure where records in each column are of the same mode and all columns are of the same length. In R, like Excel, it's typical to assign each column a label or name.

We can make a data frame from scratch with the data.frame() function. Let's build and then print a data frame called roster:

```
roster <- data.frame(
    name = c('Jack', 'Jill', 'Billy', 'Susie', 'Johnny'),
    height = c(72, 65, 68, 69, 66),
    injured = c(FALSE, TRUE, FALSE, FALSE, TRUE))

roster
#>      name height injured
#> 1    Jack     72   FALSE
#> 2    Jill     65    TRUE
#> 3   Billy     68   FALSE
#> 4   Susie     69   FALSE
#> 5  Johnny     66    TRUE
```

We've used the c() function before to combine elements into a vector. And indeed, a data frame can be thought of as a *collection of vectors* of equal length. At three variables and five observations, roster is a pretty miniscule data frame. Fortunately, a data frame doesn't always have to be built from scratch like this. For instance, R comes installed with many datasets. You can view a listing of them with this function:

```
data()
```

A menu labeled "R data sets" will appear as a new window in your scripts pane. Many, but not all, of these datasets are structured as data frames. For example, you may have encountered the famous *iris* dataset before; this is available out of the box in R.

Just like with any object, it's possible to print *iris*; however, this will quickly overwhelm your console with 150 rows of data. (Imagine the problem compounded to thousands or millions of rows.) It's more common instead to print just the first few rows with the head() function:

```
head(iris)
#> Sepal.Length Sepal.Width Petal.Length Petal.Width Species
#> 1          5.1         3.5          1.4         0.2  setosa
#> 2          4.9         3.0          1.4         0.2  setosa
#> 3          4.7         3.2          1.3         0.2  setosa
#> 4          4.6         3.1          1.5         0.2  setosa
#> 5          5.0         3.6          1.4         0.2  setosa
#> 6          5.4         3.9          1.7         0.4  setosa
```

We can confirm that iris is indeed a data frame with is.data.frame():

```
is.data.frame(iris)
#> [1] TRUE
```

Another way to get to know our new dataset besides printing it is with the `str()` function:

```
str(iris)
#> 'data.frame':        150 obs. of  5 variables:
#> $ Sepal.Length: num  5.1 4.9 4.7 4.6 5 5.4 4.6 5 4.4 4.9 ...
#> $ Sepal.Width : num  3.5 3 3.2 3.1 3.6 3.9 3.4 3.4 2.9 3.1 ...
#> $ Petal.Length: num  1.4 1.4 1.3 1.5 1.4 1.7 1.4 1.5 1.4 1.5 ...
#> $ Petal.Width : num  0.2 0.2 0.2 0.2 0.2 0.4 0.3 0.2 0.2 0.1 ...
#> $ Species     : Factor w/ 3 levels "setosa","versicolor",..: 1 1 1 1 1 1 ...
```

The output returns the size of the data frame and some information about its columns. You'll see that four of them are numeric. The last, *Species*, is a *factor*. Factors are a special way to store variables that take on a limited number of values. They are especially helpful for storing *categorical* variables: in fact, you'll see that *Species* is described as having three *levels*, which is a term we've used statistically in describing categorical variables.

Though outside the scope of this book, factors carry many benefits for working with categorical variables, such as offering more memory-efficient storage. To learn more about factors, check out R's help documentation for the `factor()` function. (This can be done with the `?` operator.) The `tidyverse` also includes `forcats` as a core package to assist in working with factors.

In addition to the datasets that are preloaded with R, many packages include their own data. You can also find out about them with the `data()` function. Let's see if the `psych` package includes any datasets:

```
data(package = 'psych')
```

The "R data sets" menu will again launch in a new window; this time, an additional section called "Data sets in package `psych`" will appear. One of these datasets is called `sat.act`. To make this dataset available to our R session, we can again use the `data()` function. It's now an assigned R object that you can find in your Environment menu and use like any other object; let's confirm it's a data frame:

```
data('sat.act')
str(sat.act)
#> 'data.frame':        700 obs. of  6 variables:
#> $ gender   : int  2 2 2 1 1 1 2 1 2 2 ...
#> $ education: int  3 3 3 4 2 5 5 3 4 5 ...
#> $ age      : int  19 23 20 27 33 26 30 19 23 40 ...
#> $ ACT      : int  24 35 21 26 31 28 36 22 22 35 ...
#> $ SATV     : int  500 600 480 550 600 640 610 520 400 730 ...
#> $ SATQ     : int  500 500 470 520 550 640 500 560 600 800 ...
```

Importing Data in R

When working in Excel, it's common to store, analyze, and present data all within the same workbook. By contrast, it's uncommon to store data from inside an R script. Generally, data will be imported from external sources, ranging from text files and databases to web pages and application programming interfaces (APIs) to images and audio, and only then analyzed in R. Results of the analysis are then frequently exported to still different sources. Let's start this process by reading data from, not surprisingly, Excel workbooks (file extension *.xlsx*), and comma-separated value files (file extension *.csv*).

Base R Versus the tidyverse

In Chapter 6 you learned about the relationship between base R and R packages. Although packages can assist in doing things that would be quite difficult to do in base R, they sometimes offer alternative ways to do the same thing. For example, base R does include functions for reading *.csv* files (but not Excel files). It also includes options for plotting. We'll be using features of the tidyverse for these and other data needs. Depending on what you're looking to do, there's nothing wrong with the base R counterparts. I've decided to focus on tidyverse tools here because its syntax is more intelligible to Excel users.

To import data in R, it's important to understand how file paths and directories work. Each time you use the program, you're working from a "home base" on your computer, or a *working directory*. Any files you refer to from R, such as when you import a dataset, are assumed to be located relative to that working directory. The getwd() function prints the working directory's file path. If you are on Windows, you will see a result similar to this:

```
getwd()
#> [1] "C:/Users/User/Documents"
```

For Mac, it will look something like this:

```
getwd()
#> [1] "/Users/user"
```

R has a global default working directory, which is the same at each session startup. I'm assuming that you are running files from a downloaded or cloned copy of the book's companion repository, and that you are also working from an R script in that same folder. In that case, you're best off setting the working directory to this folder, which can be done with the setwd() function. If you're not used to working with file paths, it can be tricky to fill this out correctly; fortunately, RStudio includes a menu-driven approach for doing it.

To change your working directory to the same folder as your current R script, go to Session → Set Working Directory → To Source File Location. You should see the results of the setwd() function appear in the console. Try running getwd() again; you'll see that you are now in a different working directory.

Now that we've established the working directory, let's practice interacting with files relative to that directory. I have placed a *test-file.csv* file in the main folder of the book repository. We can use the file.exists() function to check whether we can successfully locate it:

```
file.exists('test-file.csv')
#> [1] TRUE
```

I have also placed a copy of this file in the *test-folder* subfolder of the repository. This time, we'll need to specify which subfolder to look in:

```
file.exists('test-folder/test-file.csv')
#> [1] TRUE
```

What happens if we need to go *up* a folder? Try placing a copy of *test-file* in whatever folder is one above your current directory. We can use .. to tell R to look one folder up:

```
file.exists('../test-file.csv')
#> [1] TRUE
```

RStudio Projects

In the book repository, you will find a file called *aia-book.Rproj*. This is an RStudio project file. A project is a great way to preserve your work; for example, the project will maintain the configuration of windows and files that you had open in RStudio when you left. In addition, the project will automatically set your working directory to the *project* directory, so that you won't need a hardcoded setwd() for each script. When you work with R in this repository, then, consider doing so via the *.Rproj* file. You can then open any file via the Files pane in the lower right pane of RStudio.

Now that you have the hang of locating files in R, let's actually read some in. The book repository contains a *datasets* folder (*https://oreil.ly/wtneb*), under which is a *star* subfolder. This contains, among other things, two files: *districts.csv* and *star.xlsx*.

To read in *.csv* files, we can use the read_csv() function from readr. This package is part of the tidyverse collection, so we don't need to install or load anything new. We will pass the location of the file into the function. (Do you see now why understanding working directories and file paths was helpful?)

```
read_csv('datasets/star/districts.csv')
#>-- Column specification --------------------------
```

```
#> cols(
#>   schidkn = col_double(),
#>   school_name = col_character(),
#>   county = col_character()
#> )
#>
#> # A tibble: 89 x 3
#>    schidkn school_name     county
#>      <dbl> <chr>           <chr>
#> 1        1 Rosalia         New Liberty
#> 2        2 Montgomeryville Topton
#> 3        3 Davy            Wahpeton
#> 4        4 Steelton        Palestine
#> 5        5 Bonifay         Reddell
#> 6        6 Tolchester      Sattley
#> 7        7 Cahokia         Sattley
#> 8        8 Plattsmouth     Sugar Mountain
#> 9        9 Bainbridge      Manteca
#>10       10 Bull Run        Manteca
#> # ... with 79 more rows
```

RStudio's Import Dataset Wizard

If you are struggling to import a dataset, try RStudio's menu-driven data importer by heading to File → Import Dataset. You will be presented with a series of options to walk you through the process, including the ability to navigate to the source file via your computer's file explorer.

This results in a fair amount of output. First, our columns are specified, and we're told which functions were used to parse the data into R. Next, the first few rows of the data are listed, as a *tibble*. This is a modernized take on the data frame. It's still a data frame, and behaves mostly like a data frame, with some modifications to make it easier to work with, especially within the tidyverse.

Although we were able to read our data into R, we won't be able to do much with it unless we assign it to an object:

```
districts <- read_csv('datasets/star/districts.csv')
```

Among its many benefits, one nice thing about the tibble is we can print it without having to worry about overwhelming the console output; the first 10 rows only are printed:

```
districts
#> # A tibble: 89 x 3
#>    schidkn school_name     county
#>      <dbl> <chr>           <chr>
#> 1        1 Rosalia         New Liberty
#> 2        2 Montgomeryville Topton
```

```
#> 3       3 Davy          Wahpeton
#> 4       4 Steelton      Palestine
#> 5       5 Bonifay       Reddell
#> 6       6 Tolchester    Sattley
#> 7       7 Cahokia       Sattley
#> 8       8 Plattsmouth   Sugar Mountain
#> 9       9 Bainbridge    Manteca
#> 10      10 Bull Run      Manteca
#> # ... with 79 more rows
```

readr does not include a way to import Excel workbooks; we will instead use the readxl package. While it is part of the tidyverse, this package does not load with the core suite of packages like readr does, which is why we imported it separately at the beginning of the chapter.

We'll use the read_xlsx() function to similarly import *star.xlsx* as a tibble:

```
star <- read_xlsx('datasets/star/star.xlsx')
head(star)
#> # A tibble: 6 x 8
#>   tmathssk treadssk classk        totexpk sex   freelunk race  schidkn
#>      <dbl>    <dbl> <chr>           <dbl> <chr> <chr>    <chr>   <dbl>
#> 1      473      447 small.class         7 girl  no       white      63
#> 2      536      450 small.class        21 girl  no       black      20
#> 3      463      439 regular.wit~        0 boy   yes      black      19
#> 4      559      448 regular            16 boy   no       white      69
#> 5      489      447 small.class         5 boy   yes      white      79
#> 6      454      431 regular             8 boy   yes      white       5
```

There's more you can do with readxl, such as reading in *.xls* or *.xlsm* files and reading in specific worksheets or ranges of a workbook. To learn more, check out the package's documentation (*https://oreil.ly/kuZPE*).

Exploring a Data Frame

Earlier you learned about head() and str() to size up a data frame. Here are a few more helpful functions. First, View() is a function from RStudio whose output will be very welcome to you as an Excel user:

```
View(star)
```

After calling this function, a spreadsheet-like viewer will appear in a new window in your Scripts pane. You can sort, filter, and explore your dataset much like you would in Excel. However, as the function implies, it's for viewing *only*. You cannot make changes to the data frame from this window.

The glimpse() function is another way to print several records of the data frame, along with its column names and types. This function comes from dplyr, which is part of the tidyverse. We will lean heavily on dplyr in later chapters to manipulate data.

```
glimpse(star)
#> Rows: 5,748
#> Columns: 8
#> $ tmathssk <dbl> 473, 536, 463, 559, 489,...
#> $ treadssk <dbl> 447, 450, 439, 448, 447,...
#> $ classk   <chr> "small.class", "small.cl...
#> $ totexpk  <dbl> 7, 21, 0, 16, 5, 8, 17, ...
#> $ sex      <chr> "girl", "girl", "boy", "...
#> $ freelunk <chr> "no", "no", "yes", "no",...
#> $ race     <chr> "white", "black", "black...
#> $ schidkn  <dbl> 63, 20, 19, 69, 79, 5, 1...
```

There's also the summary() function from base R, which produces summaries of various R objects. When a data frame is passed into summary(), some basic descriptive statistics are provided:

```
summary(star)
#>     tmathssk        treadssk         classk             totexpk
#>  Min.   :320.0   Min.   :315.0   Length:5748        Min.   : 0.000
#>  1st Qu.:454.0   1st Qu.:414.0   Class :character   1st Qu.: 5.000
#>  Median :484.0   Median :433.0   Mode  :character   Median : 9.000
#>  Mean   :485.6   Mean   :436.7                       Mean   : 9.307
#>  3rd Qu.:513.0   3rd Qu.:453.0                       3rd Qu.:13.000
#>  Max.   :626.0   Max.   :627.0                       Max.   :27.000
#>      sex              freelunk            race
#>  Length:5748        Length:5748        Length:5748
#>  Class :character   Class :character   Class :character
#>  Mode  :character   Mode  :character   Mode  :character
#>     schidkn
#>  Min.   : 1.00
#>  1st Qu.:20.00
#>  Median :39.00
#>  Mean   :39.84
#>  3rd Qu.:60.00
#>  Max.   :80.00
```

Many other packages include their own version of descriptive statistics; one of my favorite is the describe() function from psych:

```
describe(star)
#>           vars    n   mean    sd median trimmed   mad min max range  skew
#> tmathssk     1 5748 485.65 47.77    484  483.20 44.48 320 626   306  0.47
#> treadssk     2 5748 436.74 31.77    433  433.80 28.17 315 627   312  1.34
#> classk*      3 5748   1.95  0.80      2    1.94  1.48   1   3     2  0.08
#> totexpk      4 5748   9.31  5.77      9    9.00  5.93   0  27    27  0.42
#> sex*         5 5748   1.49  0.50      1    1.48  0.00   1   2     1  0.06
#> freelunk*    6 5748   1.48  0.50      1    1.48  0.00   1   2     1  0.07
#> race*        7 5748   2.35  0.93      3    2.44  0.00   1   3     2 -0.75
#> schidkn      8 5748  39.84 22.96     39   39.76 29.65   1  80    79  0.04
#>           kurtosis    se
```

```
#> tmathssk      0.29 0.63
#> treadssk      3.83 0.42
#> classk*      -1.45 0.01
#> totexpk      -0.21 0.08
#> sex*         -2.00 0.01
#> freelunk*    -2.00 0.01
#> race*        -1.43 0.01
#> schidkn      -1.23 0.30
```

If you're not familiar with all of these descriptive statistics, you know what to do: *check the function's documentation.*

Indexing and Subsetting Data Frames

Earlier in this section we created a small data frame roster containing the names and heights of four individuals. Let's demonstrate some basic data frame manipulation techniques with this object.

In Excel, you can use the INDEX() function to refer to both the row and column positions of a table, as shown in Figure 7-4:

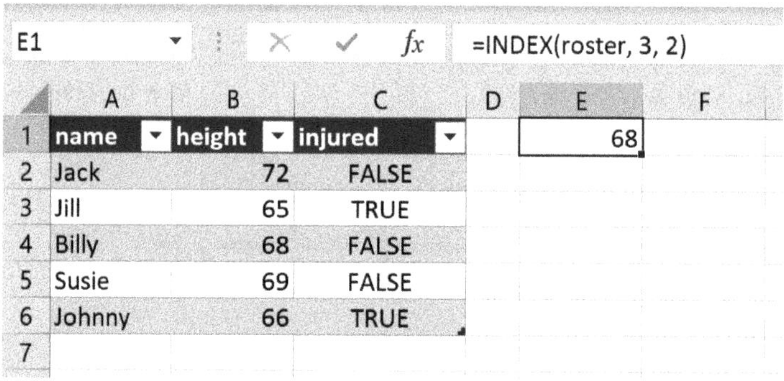

Figure 7-4. The INDEX() function on an Excel table

This will work similarly in R. We'll use the same bracket notation as we to with index vectors, but this time we'll refer to both the row and column position:

```
# Third row, second column of data frame
roster[3, 2]
#> [1] 68
```

Again, we can use the : operator to retrieve all elements within a given range:

```
# Second through fourth rows, first through third columns
roster[2:4, 1:3]
#>     name height injured
#> 2 Jill      65    TRUE
```

```
#> 3 Billy    68    FALSE
#> 4 Susie    69    FALSE
```

It's also possible to select an entire row or column by leaving its index blank, or to use the c() function to subset nonconsecutive elements:

```
# Second and third rows only
roster[2:3,]
#>    name height injured
#> 2  Jill    65    TRUE
#> 3 Billy    68    FALSE

# First and third columns only
roster[, c(1,3)]
#>     name injured
#> 1   Jack   FALSE
#> 2   Jill    TRUE
#> 3  Billy   FALSE
#> 4  Susie   FALSE
#> 5 Johnny    TRUE
```

If we just want to access one column of the data frame, we can use the $ operator. Interestingly, this results in a *vector*:

```
roster$height
#> [1] 72 65 68 69 66
is.vector(roster$height)
#> [1] TRUE
```

This confirms that a data frame is indeed a list of vectors of equal length.

Other Data Structures in R

We've focused on R's vector and data frame structures as they are equivalents to Excel's ranges and tables and those you're most likely to work with for data analysis. There are, however, several other data structures in base R such as matrices and lists. To learn more about these structures and how they relate to vectors and data frames, check out Hadley Wickham's *Advanced R*, 2nd edition (Chapman & Hall).

Writing Data Frames

As mentioned earlier, it's typical to read data into R, operate on it, and then export the results elsewhere. To write a data frame to a *.csv* file, you can use the write_csv() function from readr:

```
# Write roster data frame to csv
write_csv(roster, 'output/roster-output-r.csv')
```

If you have the working directory set to the book's companion repository, you should find this file waiting for you in the *output* folder.

Unfortunately, the readxl package does not include a function to write data to an Excel workbook. We can, however, use writexl and its write_xlsx() function:

```
# Write roster data frame to csv
write_xlsx(roster, 'output/roster-output-r.xlsx')
```

Conclusion

In this chapter, you progressed from single-element objects, to larger vectors, and finally to data frames. While we'll be working with data frames for the remainder of the book, it's helpful to keep in mind that they are collections of vectors and behave largely in the same way. Coming up, you will learn how to analyze, visualize, and ultimately test relationships in R data frames.

Exercises

Do the following exercises to test your knowledge of data structures in R:

1. Create a character vector of five elements, and then access the first and fourth elements of this vector.

2. Create two vectors x and y of length 4, one containing numeric and the other logical values. Multiply them and pass the result to z. What is the result?

3. Download the nycflights13 package from CRAN. How many datasets are included with this package?

   • One of these datasets is called airports. Print the first few rows of this data frame as well as the descriptive statistics.

   • Another is called weather. Find the 10th through 12th rows and the 4th through 7th columns of this data frame. Write the results to a *.csv* file and an Excel workbook.

## CHAPTER 8

Data Manipulation and Visualization in R

American statistician Ronald Thisted once quipped: "Raw data, like raw potatoes, usually require cleaning before use." Data manipulation takes time, and you've felt the pain if you've ever done the following:

- Select, drop, or create calculated columns
- Sort or filter rows
- Group by and summarize categories
- Join multiple datasets by a common field

Chances are, you've done all of these in Excel...*a lot*, and you've probably dug into celebrated features like VLOOKUP( ) and PivotTables to accomplish them. In this chapter, you'll learn the R equivalents of these techniques, particularly with the help of dplyr.

Data manipulation often goes hand in hand with visualization: as mentioned, humans are remarkably adept at visually processing information, so it's a great way to size up a dataset. You'll learn how to visualize data using the gorgeous ggplot2 package, which like dplyr is part of the tidyverse. This will put you on solid footing to explore and test relationships in data using R, which will be covered in Chapter 9. Let's get started by calling in the relevant packages. We'll also be using the *star* dataset from the book's companion repository (*https://oreil.ly/lmZb7*) in this chapter, so we can import it now:

```
library(tidyverse)
library(readxl)

star <- read_excel('datasets/star/star.xlsx')
head(star)
#> # A tibble: 6 x 8
```

```
#>    tmathssk treadssk classk              totexpk sex   freelunk race  schidkn
#>       <dbl>    <dbl> <chr>                  <dbl> <chr> <chr>    <chr>   <dbl>
#> 1       473      447 small.class                7 girl  no       white      63
#> 2       536      450 small.class               21 girl  no       black      20
#> 3       463      439 regular.with.aide          0 boy   yes      black      19
#> 4       559      448 regular                   16 boy   no       white      69
#> 5       489      447 small.class                5 boy   yes      white      79
#> 6       454      431 regular                    8 boy   yes      white       5
```

Data Manipulation with dplyr

dplyr is a popular package built to manipulate tabular data structures. Its many functions, or *verbs*, work similarly and can be easily used together. Table 8-1 lists some common dplyr functions and their uses; this chapter covers each of these.

Table 8-1. Frequently used verbs of dplyr

Function	What it does
select()	Selects given columns
mutate()	Creates new columns based on existing columns
rename()	Renames given columns
arrange()	Reorders rows given criteria
filter()	Selects rows given criteria
group_by()	Groups rows by given columns
summarize()	Aggregates values for each group
left_join()	Joins matching records from Table B to Table A; result is NA if no match found in Table B

For the sake of brevity, I won't cover all of the functions of dplyr or even all the ways to use the functions that we do cover. To learn more about the package, check out *R for Data Science* by Hadley Wickham and Garrett Grolemund (O'Reilly). You can also access a helpful cheat sheet summarizing how the many functions of dplyr work together by navigating in RStudio to Help → Cheatsheets → Data Transformation with dplyr.

Column-Wise Operations

Selecting and dropping columns in Excel often requires hiding or deleting them. This can be difficult to audit or reproduce, because hidden columns are easily overlooked, and deleted columns aren't easily recovered. The select() function can be used to choose given columns from a data frame in R. For select(), as with each of these functions, the first argument will be which data frame to work with. Additional arguments are then provided to manipulate the data in that data frame. For example, we can select *tmathssk*, *treadssk*, and *schidkin* from star like this:

```
select(star, tmathssk, treadssk, schidkn)
#> # A tibble: 5,748 x 3
#>    tmathssk treadssk schidkn
#>       <dbl>    <dbl>   <dbl>
#>  1      473      447      63
#>  2      536      450      20
#>  3      463      439      19
#>  4      559      448      69
#>  5      489      447      79
#>  6      454      431       5
#>  7      423      395      16
#>  8      500      451      56
#>  9      439      478      11
#> 10      528      455      66
#> # ... with 5,738 more rows
```

We can also use the - operator with select() to *drop* given columns:

```
select(star, -tmathssk, -treadssk, -schidkn)
#> # A tibble: 5,748 x 5
#>    classk            totexpk sex   freelunk race
#>    <chr>               <dbl> <chr> <chr>    <chr>
#>  1 small.class             7 girl  no       white
#>  2 small.class            21 girl  no       black
#>  3 regular.with.aide       0 boy   yes      black
#>  4 regular                16 boy   no       white
#>  5 small.class             5 boy   yes      white
#>  6 regular                 8 boy   yes      white
#>  7 regular.with.aide      17 girl  yes      black
#>  8 regular                 3 girl  no       white
#>  9 small.class            11 girl  no       black
#> 10 small.class            10 girl  no       white
```

A more elegant alternative here is to pass all unwanted columns into a vector, *then* drop it:

```
select(star, -c(tmathssk, treadssk, schidkn))
#> # A tibble: 5,748 x 5
#>    classk            totexpk sex   freelunk race
#>    <chr>               <dbl> <chr> <chr>    <chr>
#>  1 small.class             7 girl  no       white
#>  2 small.class            21 girl  no       black
#>  3 regular.with.aide       0 boy   yes      black
#>  4 regular                16 boy   no       white
#>  5 small.class             5 boy   yes      white
#>  6 regular                 8 boy   yes      white
#>  7 regular.with.aide      17 girl  yes      black
#>  8 regular                 3 girl  no       white
#>  9 small.class            11 girl  no       black
#> 10 small.class            10 girl  no       white
```

Keep in mind that in the previous examples, we've just been calling functions: we didn't actually assign the output to an object.

One more bit of shorthand for select() is to use the : operator to select everything between two columns, inclusive. This time, I will assign the results of selecting everything from *tmathssk* to *totexpk* back to star:

```
star <- select(star, tmathssk:totexpk)
head(star)
#> # A tibble: 6 x 4
#>   tmathssk treadssk classk              totexpk
#>      <dbl>    <dbl> <chr>                 <dbl>
#> 1      473      447 small.class               7
#> 2      536      450 small.class              21
#> 3      463      439 regular.with.aide         0
#> 4      559      448 regular                  16
#> 5      489      447 small.class               5
#> 6      454      431 regular                   8
```

You've likely created calculated columns in Excel; mutate() will do the same in R. Let's create a column *new_column* of combined reading and math scores. With mutate(), we'll provide the name of the new column *first*, then an equal sign, and finally the calculation to use. We can refer to other columns as part of the formula:

```
star <- mutate(star, new_column = tmathssk + treadssk)
head(star)
#> # A tibble: 6 x 5
#>   tmathssk treadssk classk              totexpk new_column
#>      <dbl>    <dbl> <chr>                 <dbl>      <dbl>
#> 1      473      447 small.class               7        920
#> 2      536      450 small.class              21        986
#> 3      463      439 regular.with.aide         0        902
#> 4      559      448 regular                  16       1007
#> 5      489      447 small.class               5        936
#> 6      454      431 regular                   8        885
```

mutate() makes it easy to derive relatively more complex calculated columns such as logarithmic transformations or lagged variables; check out the help documentation for more.

new_column isn't a particularly helpful name for total score. Fortunately, the rename() function does what it sounds like it would. We'll specify what to name the new column in place of the old:

```
star <- rename(star, ttl_score = new_column)
head(star)
#> # A tibble: 6 x 5
#>   tmathssk treadssk classk              totexpk ttl_score
#>      <dbl>    <dbl> <chr>                 <dbl>     <dbl>
#> 1      473      447 small.class               7       920
#> 2      536      450 small.class              21       986
#> 3      463      439 regular.with.aide         0       902
#> 4      559      448 regular                  16      1007
```

```
#> 5      489      447 small.class        5      936
#> 6      454      431 regular            8      885
```

Row-Wise Operations

Thus far we've been operating on *columns*. Now let's focus on *rows*; specifically sorting and filtering. In Excel, we can sort by multiple columns with the Custom Sort menu. Say for example we wanted to sort this data frame by *classk*, then *treadssk*, both ascending. Our menu in Excel to do this would look like Figure 8-1.

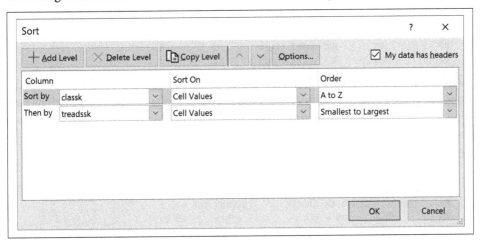

Figure 8-1. The Custom Sort menu in Excel

We can replicate this in dplyr by using the arrange() function, including each column in the order in which we want the data frame sorted:

```
arrange(star, classk, treadssk)
#> # A tibble: 5,748 x 5
#>     tmathssk treadssk classk   totexpk ttl_score
#>        <dbl>    <dbl> <chr>      <dbl>     <dbl>
#> 1        320      315 regular        3       635
#> 2        365      346 regular        0       711
#> 3        384      358 regular       20       742
#> 4        384      358 regular        3       742
#> 5        320      360 regular        6       680
#> 6        423      376 regular       13       799
#> 7        418      378 regular       13       796
#> 8        392      378 regular       13       770
#> 9        392      378 regular        3       770
#> 10       399      380 regular        6       779
#> # ... with 5,738 more rows
```

We can pass the desc() function to a column if we'd like that column to be sorted descendingly.

```
# Sort by classk descending, treadssk ascending
arrange(star, desc(classk), treadssk)
#> # A tibble: 5,748 x 5
#>    tmathssk treadssk classk       totexpk ttl_score
#>       <dbl>    <dbl> <chr>          <dbl>     <dbl>
#>  1      412      370 small.class       15       782
#>  2      434      376 small.class       11       810
#>  3      423      378 small.class        6       801
#>  4      405      378 small.class        8       783
#>  5      384      380 small.class       19       764
#>  6      405      380 small.class       15       785
#>  7      439      382 small.class        8       821
#>  8      384      384 small.class       10       768
#>  9      405      384 small.class        8       789
#> 10      423      384 small.class       21       807
```

Excel tables include helpful drop-down menus to filter any column by given conditions. To filter a data frame in R, we'll use the aptly named filter() function. Let's filter star to keep only the records where classk is equal to small.class. Remember that because we are checking for equality rather than assigning an object, we'll have to use == and not = here:

```
filter(star, classk == 'small.class')
#> # A tibble: 1,733 x 5
#>    tmathssk treadssk classk       totexpk ttl_score
#>       <dbl>    <dbl> <chr>          <dbl>     <dbl>
#>  1      473      447 small.class        7       920
#>  2      536      450 small.class       21       986
#>  3      489      447 small.class        5       936
#>  4      439      478 small.class       11       917
#>  5      528      455 small.class       10       983
#>  6      559      474 small.class        0      1033
#>  7      494      424 small.class        6       918
#>  8      478      422 small.class        8       900
#>  9      602      456 small.class       14      1058
#> 10      439      418 small.class        8       857
#> # ... with 1,723 more rows
```

We can see from the tibble output that our filter() operation *only* affected the number of rows, *not* the columns. Now we'll find the records where treadssk is at least 500:

```
filter(star, treadssk >= 500)
#> # A tibble: 233 x 5
#>    tmathssk treadssk classk            totexpk ttl_score
#>       <dbl>    <dbl> <chr>               <dbl>     <dbl>
#>  1      559      522 regular                 8      1081
#>  2      536      507 regular.with.aide       3      1043
#>  3      547      565 regular.with.aide       9      1112
#>  4      513      503 small.class             7      1016
#>  5      559      605 regular.with.aide       5      1164
#>  6      559      554 regular                14      1113
```

```
#>  7     559     503 regular          10     1062
#>  8     602     518 regular          12     1120
#>  9     536     580 small.class      12     1116
#> 10     626     510 small.class      14     1136
#> # ... with 223 more rows
```

It's possible to filter by multiple conditions using the & operator for "and" along with the | operator for "or." Let's combine our two criteria from before with &:

```
# Get records where classk is small.class and
# treadssk is at least 500
filter(star, classk == 'small.class' & treadssk >= 500)
#> # A tibble: 84 x 5
#>    tmathssk treadssk classk      totexpk ttl_score
#>       <dbl>    <dbl> <chr>         <dbl>     <dbl>
#>  1      513      503 small.class       7      1016
#>  2      536      580 small.class      12      1116
#>  3      626      510 small.class      14      1136
#>  4      602      518 small.class       3      1120
#>  5      626      565 small.class      14      1191
#>  6      602      503 small.class      14      1105
#>  7      626      538 small.class      13      1164
#>  8      500      580 small.class       8      1080
#>  9      489      565 small.class      19      1054
#> 10      576      545 small.class      19      1121
#> # ... with 74 more rows
```

Aggregating and Joining Data

I like to call PivotTables "the WD-40 of Excel" because they allow us to get our data "spinning" in different directions for easy analysis. For example, let's recreate the PivotTable in Figure 8-2 showing the average math score by class size from the *star* dataset:

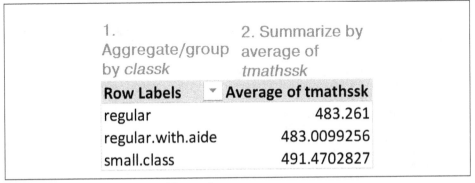

Figure 8-2. How Excel PivotTables work

As Figure 8-2 calls out, there are two elements to this PivotTable. First, I aggregated our data by the variable *classk*. Then, I summarized it by taking an average of *tmathssk.* In R, these are discrete steps, using different `dplyr` functions. First, we'll aggregate the data using `group_by()`. Our output includes a line, `# Groups: classk [3]`, indicating that `star_grouped` is split into three groups with the `classk` variable:

```
star_grouped <- group_by(star, classk)
head(star_grouped)
#> # A tibble: 6 x 5
#> # Groups:   classk [3]
#>    tmathssk treadssk classk            totexpk ttl_score
#>       <dbl>    <dbl> <chr>               <dbl>     <dbl>
#> 1       473      447 small.class             7       920
#> 2       536      450 small.class            21       986
#> 3       463      439 regular.with.aide       0       902
#> 4       559      448 regular                16      1007
#> 5       489      447 small.class             5       936
#> 6       454      431 regular                 8       885
```

We've *grouped* our data by one variable; now let's *summarize* it by another with the `summarize()` function (`summarise()` also works). Here we'll specify what to name the resulting column, and how to calculate it. Table 8-2 lists some common aggregation functions.

Table 8-2. Helpful aggregation functions for `dplyr`

Function	Aggregation type
sum()	Sum
n()	Count values
mean()	Average
max()	Highest value
min()	Lowest value
sd()	Standard deviation

We can get the average math score by class size by running `summarize()` on our grouped data frame:

```
summarize(star_grouped, avg_math = mean(tmathssk))
#> `summarise()` ungrouping output (override with `.groups` argument)
#> # A tibble: 3 x 2
#>   classk            avg_math
#>   <chr>                <dbl>
#> 1 regular               483.
#> 2 regular.with.aide     483.
#> 3 small.class           491.
```

The `summarise()` ungrouping output error is a warning that you've ungrouped the grouped tibble by aggregating it. Minus some formatting differences, we have the same results as Figure 8-2.

If PivotTables are the WD-40 of Excel, then VLOOKUP() is the duct tape, allowing us to easily combine data from multiple sources. In our original *star* dataset, *schidkin* is a school district indicator. We dropped this column earlier in this chapter, so let's read it in again. But what if in addition to the indicator number we actually wanted to know the *names* of these districts? Fortunately, *districts.csv* in the book repository has this information, so let's read both in and come up with a strategy for combining them:

```
star <- read_excel('datasets/star/star.xlsx')
head(star)
#> # A tibble: 6 x 8
#>   tmathssk treadssk classk          totexpk sex   freelunk race  schidkn
#>      <dbl>    <dbl> <chr>             <dbl> <chr> <chr>    <chr>   <dbl>
#> 1      473      447 small.class           7 girl  no       white      63
#> 2      536      450 small.class          21 girl  no       black      20
#> 3      463      439 regular.with.aide     0 boy   yes      black      19
#> 4      559      448 regular              16 boy   no       white      69
#> 5      489      447 small.class           5 boy   yes      white      79
#> 6      454      431 regular               8 boy   yes      white       5

districts <- read_csv('datasets/star/districts.csv')

#> -- Column specification -------------------------------------------------
#> cols(
#>   schidkn = col_double(),
#>   school_name = col_character(),
#>   county = col_character()
#> )

head(districts)
#> # A tibble: 6 x 3
#>   schidkn school_name     county
#>     <dbl> <chr>           <chr>
#> 1       1 Rosalia         New Liberty
#> 2       2 Montgomeryville Topton
#> 3       3 Davy            Wahpeton
#> 4       4 Steelton        Palestine
#> 5       6 Tolchester      Sattley
#> 6       7 Cahokia         Sattley
```

It appears that what's needed is like a VLOOKUP(): we want to "read in" the *school_name* (and possibly the *county*) variables from *districts* into *star*, given the shared *schidkn* variable. To do this in R, we'll use the methodology of *joins*, which comes from relational databases, a topic that was touched on in Chapter 5. Closest to a VLOOKUP() is the left outer join, which can be done in dplyr with the left_join() function. We'll provide the "base" table first (*star*) and then the "lookup" table (*districts*). The function will look for and return a match in *districts* for every record in *star*, or return NA if no match is found. I will keep only some columns from *star* for less overwhelming console output:

```
# Left outer join star on districts
left_join(select(star, schidkn, tmathssk, treadssk), districts)
#> Joining, by = "schidkn"
#> # A tibble: 5,748 x 5
#>    schidkn tmathssk treadssk school_name    county
#>      <dbl>    <dbl>    <dbl> <chr>          <chr>
#>  1      63      473      447 Ridgeville     New Liberty
#>  2      20      536      450 South Heights  Selmont
#>  3      19      463      439 Bunnlevel      Sattley
#>  4      69      559      448 Hokah          Gallipolis
#>  5      79      489      447 Lake Mathews   Sugar Mountain
#>  6       5      454      431 NA             NA
#>  7      16      423      395 Calimesa       Selmont
#>  8      56      500      451 Lincoln Heights Topton
#>  9      11      439      478 Moose Lake     Imbery
#> 10      66      528      455 Siglerville    Summit Hill
#> # ... with 5,738 more rows
```

left_join() is pretty smart: it knew to join on schidkn, and it "looked up" not just *school_name* but also *county*. To learn more about joining data, check out the help documentation.

In R, missing observations are represented as the special value NA. For example, it appears that no match was found for the name of district 5. In a VLOOKUP(), this would result in an #N/A error. An NA does *not* mean that an observation is equal to zero, only that its value is missing. You may see other special values such as NaN or NULL while programming R; to learn more about them, launch the help documentation.

dplyr and the Power of the Pipe (%>%)

As you're beginning to see, dplyr functions are powerful and rather intuitive to any-one who's worked with data, including in Excel. And as anyone who's worked with data knows, it's rare to prepare the data as needed in just one step. Take, for example, a typical data analysis task that you might want to do with *star*:

> Find the average reading score by class type, sorted high to low.

Knowing what we do about working with data, we can break this into three distinct steps:

1. Group our data by class type.
2. Find the average reading score for each group.
3. Sort these results from high to low.

We could carry this out in dplyr doing something like the following:

```
star_grouped <- group_by(star, classk)
star_avg_reading <- summarize(star_grouped, avg_reading = mean(treadssk))
#> `summarise()` ungrouping output (override with `.groups` argument)
#>
star_avg_reading_sorted <- arrange(star_avg_reading, desc(avg_reading))
star_avg_reading_sorted
#>
#> # A tibble: 3 x 2
#>   classk            avg_reading
#>   <chr>                   <dbl>
#> 1 small.class              441.
#> 2 regular.with.aide        435.
#> 3 regular                  435.
```

This gets us to an answer, but it took quite a few steps, and it can be hard to follow along with the various functions and object names. The alternative is to link these functions together with the %>%, or pipe, operator. This allows us to pass the output of one function directly into the input of another, so we're able to avoid continuously renaming our inputs and outputs. The default keyboard shortcut for this operator is Ctrl+Shift+M for Windows, Cmd-Shift-M for Mac.

Let's re-create the previous steps, this time with the pipe operator. We'll place each function on its own line, combining them with %>%. While it's not necessary to place each step on its own line, it's often preferred for legibility. When using the pipe operator, it's also not necessary to highlight the entire code block to run it; simply place your cursor anywhere in the following selection and execute:

```
star %>%
  group_by(classk) %>%
  summarise(avg_reading = mean(treadssk)) %>%
  arrange(desc(avg_reading))
#> `summarise()` ungrouping output (override with `.groups` argument)
#> # A tibble: 3 x 2
#>   classk            avg_reading
#>   <chr>                   <dbl>
#> 1 small.class              441.
#> 2 regular.with.aide        435.
#> 3 regular                  435.
```

It can be pretty disorienting at first to no longer be explicitly including the data source as an argument in each function. But compare the last code block to the one before and you can see how much more efficient this approach can be. What's more, the pipe operator can be used with non-dplyr functions. For example, let's just assign the first few rows of the resulting operation by including head() at the end of the pipe:

```
# Average math and reading score
# for each school district
star %>%
  group_by(schidkn) %>%
```

```
    summarise(avg_read = mean(treadssk), avg_math = mean(tmathssk)) %>%
    arrange(schidkn) %>%
    head()
#> `summarise()` ungrouping output (override with `.groups` argument)
#> # A tibble: 6 x 3
#>   schidkn avg_read avg_math
#>     <dbl>    <dbl>    <dbl>
#> 1       1     444.     492.
#> 2       2     407.     451.
#> 3       3     441      491.
#> 4       4     422.     468.
#> 5       5     428.     460.
#> 6       6     428.     470.
```

Reshaping Data with tidyr

Although it's true that group_by() along with summarize() serve as a PivotTable equivalent in R, these functions can't do everything that an Excel PivotTable can do. What if, instead of just aggregating the data, you wanted to *reshape* it, or change how rows and columns are set up? For example, our *star* data frame has two separate columns for math and reading scores, *tmathssk* and *treadssk*, respectively. I would like to combine these into one column called *score*, with another called *test_type* indicating whether each observation is for math or reading. I'd also like to keep the school indicator, *schidkn*, as part of the analysis.

Figure 8-3 shows what this might look like in Excel; note that I relabeled the Values fields from *tmathssk* and *treadssk* to *math* and *reading*, respectively. If you would like to inspect this PivotTable further, it is available in the book repository as *ch-8.xlsx* (*https://oreil.ly/Kq93s*). Here I am again making use of an index column; otherwise, the PivotTable would attempt to "roll up" all values by *schidkn*.

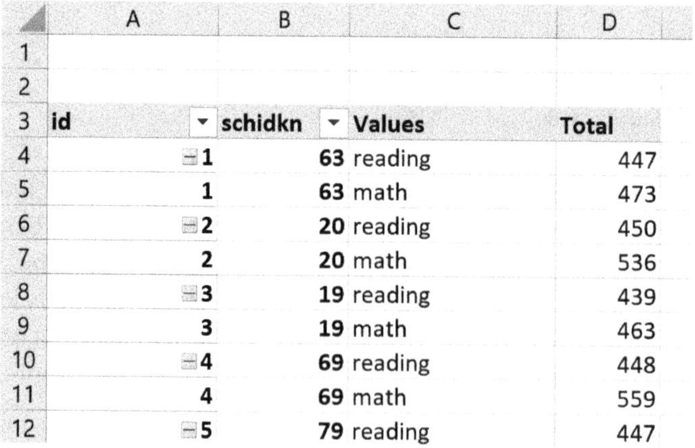

	A	B	C	D
1				
2				
3	id	schidkn	Values	Total
4	1	63	reading	447
5	1	63	math	473
6	2	20	reading	450
7	2	20	math	536
8	3	19	reading	439
9	3	19	math	463
10	4	69	reading	448
11	4	69	math	559
12	5	79	reading	447

Figure 8-3. Reshaping star in Excel

We can use `tidyr`, a core `tidyverse` package, to reshape *star*. Adding an index column will also be helpful when reshaping in R, as it was in Excel. We can make one with the `row_number()` function:

```
star_pivot <- star %>%
            select(c(schidkn, treadssk, tmathssk)) %>%
            mutate(id = row_number())
```

To reshape the data frame, we'll use `pivot_longer()` and `pivot_wider()`, both from `tidyr`. Consider in your mind's eye and in Figure 8-3 what would happen to our dataset if we consolidated scores from *tmathssk* and *treadssk* into one column. Would the dataset get longer or wider? We're adding rows here, so our dataset will get longer. To use `pivot_longer()`, we'll specify with the `cols` argument what columns to lengthen by, and use `values_to` to name the resulting column. We'll also use `names_to` to name the column indicating whether each score is math or reading:

```
star_long <- star_pivot %>%
            pivot_longer(cols = c(tmathssk, treadssk),
                        values_to = 'score', names_to = 'test_type')
head(star_long)
#> # A tibble: 6 x 4
#>    schidkn    id test_type score
#>      <dbl> <int> <chr>     <dbl>
#> 1       63     1 tmathssk    473
#> 2       63     1 treadssk    447
#> 3       20     2 tmathssk    536
#> 4       20     2 treadssk    450
#> 5       19     3 tmathssk    463
#> 6       19     3 treadssk    439
```

Great work. But is there a way to rename *tmathssk* and *treadssk* to *math* and *reading*, respectively? There is, with `recode()`, yet another helpful `dplyr` function that can be used with `mutate()`. `recode()` works a little differently than other functions in the package because we include the name of the "old" values *before* the equals sign, then the new. The `distinct()` function from `dplyr` will confirm that all rows have been named either *math* or *reading*:

```
# Rename tmathssk and treadssk as math and reading
star_long <- star_long %>%
   mutate(test_type = recode(test_type,
                     'tmathssk' = 'math', 'treadssk' = 'reading'))

distinct(star_long, test_type)
#> # A tibble: 2 x 1
#>    test_type
#>      <chr>
#> 1 math
#> 2 reading
```

Now that our data frame is lengthened, we can widen it back with `pivot_wider()`. This time, I'll specify which column has values in its rows that should be columns with `values_from`, and what the resulting columns should be named with `names_from`:

```
star_wide <- star_long %>%
            pivot_wider(values_from = 'score', names_from = 'test_type')
head(star_wide)
#> # A tibble: 6 x 4
#>    schidkn    id  math reading
#>      <dbl> <int> <dbl>   <dbl>
#> 1       63     1   473     447
#> 2       20     2   536     450
#> 3       19     3   463     439
#> 4       69     4   559     448
#> 5       79     5   489     447
#> 6        5     6   454     431
```

Reshaping data is a relatively trickier operation in R, so when in doubt, ask yourself: *am I making this data wider or longer? How would I do it in a PivotTable?* If you can logically walk through what needs to happen to achieve the desired end state, coding it will be that much easier.

Data Visualization with ggplot2

There's so much more that `dplyr` can do to help us manipulate data, but for now let's turn our attention to data visualization. Specifically, we'll focus on another `tidyverse` package, `ggplot2`. Named and modeled after the "grammar of graphics" devised by computer scientist Leland Wilkinson, `ggplot2` provides an ordered approach for constructing plots. This structure is patterned after how elements of speech come together to make a sentence, hence the "grammar" of graphics.

I'll cover some of the basic elements and plot types of `ggplot2` here. For more about the package, check out *ggplot2: Elegant Graphics for Data Analysis* by the package's original author, Hadley Wickham (Springer). You can also access a helpful cheat sheet for working with the package by navigating in RStudio to Help → Cheatsheets → Data Visualization with ggplot2. Some essential elements of `ggplot2` are found in Table 8-3. Other elements are available; for more information, check out the resources mentioned earlier.

Table 8-3. The foundational elements of ggplot2

Element	Description
data	The source data
aes	The aesthetic mappings from data to visual properties (x- and y-axes, color, size, and so forth)
geom	The type of geometric object observed in the plot (lines, bars, dots, and so forth)

Let's get started by visualizing the number of observations for each level of *classk* as a barplot. We'll start with the ggplot() function and specify the three elements from Table 8-3:

```
ggplot(data = star, ❶
          aes(x = classk)) + ❷
    geom_bar() ❸
```

❶ The data source is specified with the data argument.

❷ The aesthetic mappings from the data to the visualization are specified with the aes() function. Here we are calling for *classk* to be mapped to the x-axis of the eventual plot.

❸ We plot a geometric object based on our specified data and aesthetic mappings with the geom_bar() function. The results are shown in Figure 8-4.

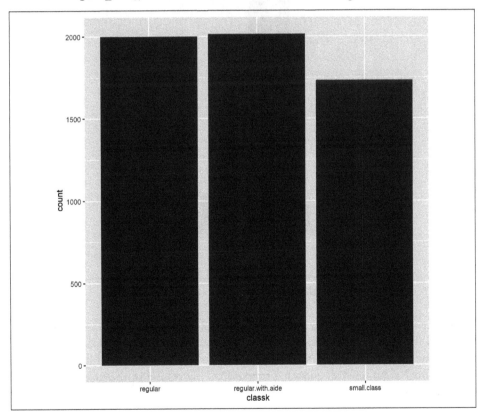

Figure 8-4. A barplot in ggplot2

Similar to the pipe operator, it's not necessary to place each layer of the plot on its own line, but it's often preferred for legibility. It's also possible to execute the entire plot by placing the cursor anywhere inside the code block and running.

Because of its modular approach, it's easy to iterate on visualizations with ggplot2. For example, we can switch our plot to a histogram of *treadssk* by changing our x mapping and plotting the results with geom_histogram(). This results in the histogram shown in Figure 8-5:

```
ggplot(data = star, aes(x = treadssk)) +
  geom_histogram()
```

```
#> `stat_bin()` using `bins = 30`. Pick better value with `binwidth`.
```

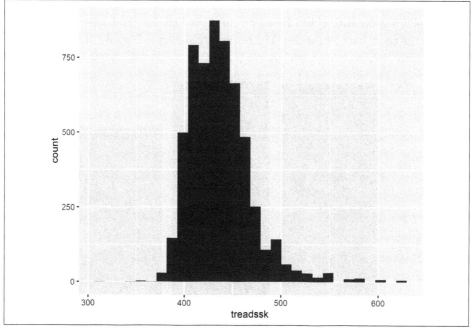

Figure 8-5. A histogram in ggplot2

There are also many ways to customize ggplot2 plots. You may have noticed, for example, that the output message for the previous plot indicated that 30 bins were used in the histogram. Let's change that number to 25 and use a pink fill with a couple of additional arguments in geom_histogram(). This results in the histogram shown in Figure 8-6:

```
ggplot(data = star, aes(x = treadssk)) +
  geom_histogram(bins = 25, fill = 'pink')
```

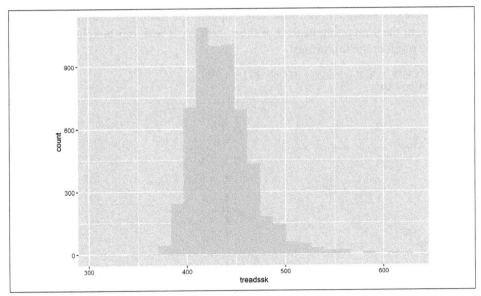

Figure 8-6. A customized histogram in `ggplot2`

Use `geom_boxplot()` to create a boxplot, as shown in Figure 8-7:

```
ggplot(data = star, aes(x = treadssk)) +
  geom_boxplot()
```

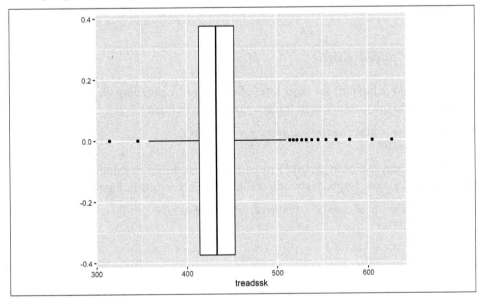

Figure 8-7. A boxplot

In any of the cases thus far, we could have "flipped" the plot by including the variable of interest in the y mapping instead of the x. Let's try it with our boxplot. Figure 8-8 shows the result of the following:

```
ggplot(data = star, aes(y = treadssk)) +
  geom_boxplot()
```

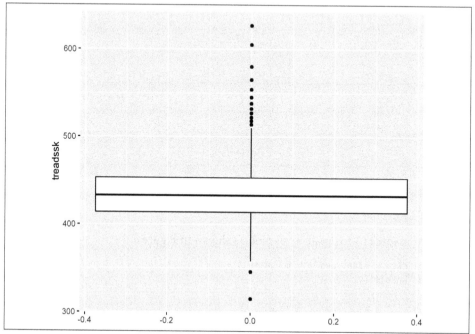

Figure 8-8. A "flipped" boxplot

Now let's make a boxplot for each level of class size by mapping *classk* to the x-axis and *treadssk* to the y, resulting in the boxplot shown in Figure 8-9:

```
ggplot(data = star, aes(x = classk, y = treadssk)) +
  geom_boxplot()
```

Similarly, we can use geom_point() to plot the relationship of *tmathssk* and *treadssk* on the x- and y-axes, respectively, as a scatterplot. This results in Figure 8-10:

```
ggplot(data = star, aes(x = tmathssk, y = treadssk)) +
  geom_point()
```

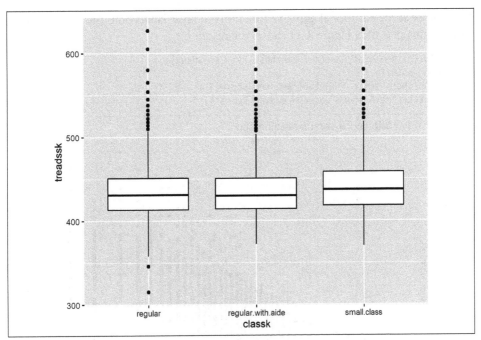

Figure 8-9. A boxplot by group

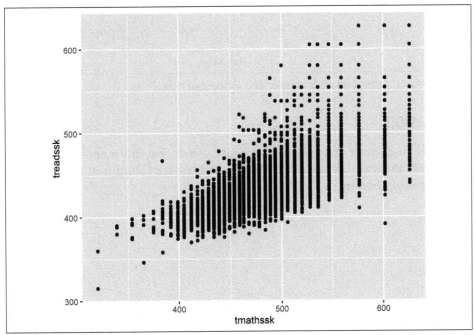

Figure 8-10. A scatterplot

We can use some additional `ggplot2` functions to layer labels onto the x- and y-axes, along with a plot title. Figure 8-11 shows the result:

```
ggplot(data = star, aes(x = tmathssk, y = treadssk)) +
  geom_point() +
  xlab('Math score') + ylab('Reading score') +
  ggtitle('Math score versus reading score')
```

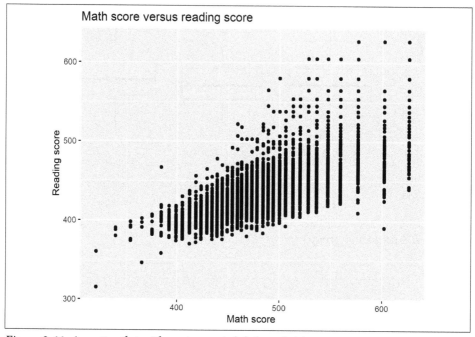

Figure 8-11. A scatterplot with custom axis labels and title

Conclusion

There's so much more that `dplyr` and `ggplot2` can do, but this is enough to get you started with the true task at hand: to explore and test relationships in data. That will be the focus of Chapter 9.

Exercises

The book repository (*https://oreil.ly/kBk3e*) has two files in the *census* subfolder of *datasets*, *census.csv* and *census-divisions.csv*. Read these into R and do the following:

1. Sort the data by region ascending, division ascending, and population descending. (You will need to combine datasets to do this.) Write the results to an Excel worksheet.
2. Drop the postal code field from your merged dataset.
3. Create a new column *density* that is a calculation of population divided by land area.
4. Visualize the relationship between land area and population for all observations in 2015.
5. Find the total population for each region in 2015.
6. Create a table containing state names and populations, with the population for each year 2010–2015 kept in an individual column.

## CHAPTER 9

Capstone: R for Data Analytics

In this chapter, we'll apply what we've learned about data analysis and visualization in R to explore and test relationships in the familiar *mpg* dataset. You'll learn a couple of new R techniques here, including how to conduct a t-test and linear regression. We'll begin by calling up the necessary packages, reading in *mpg.csv* from the *mpg* subfolder of the book repository's *datasets* folder, and selecting the columns of interest. We've not used tidymodels so far in this book, so you may need to install it.

```
library(tidyverse)
library(psych)
library(tidymodels)

# Read in the data, select only the columns we need
mpg <- read_csv('datasets/mpg/mpg.csv') %>%
  select(mpg, weight, horsepower, origin, cylinders)

#> -- Column specification ------------------------------------------
#> cols(
#>   mpg = col_double(),
#>   cylinders = col_double(),
#>   displacement = col_double(),
#>   horsepower = col_double(),
#>   weight = col_double(),
#>   acceleration = col_double(),
#>   model.year = col_double(),
#>   origin = col_character(),
#>   car.name = col_character()
#> )

head(mpg)
#> # A tibble: 6 x 5
#>     mpg weight horsepower origin cylinders
#>   <dbl>  <dbl>      <dbl> <chr>      <dbl>
#> #> 1    18   3504        130 USA            8
```

```
#> 2    15    3693       165 USA          8
#> 3    18    3436       150 USA          8
#> 4    16    3433       150 USA          8
#> 5    17    3449       140 USA          8
#> 6    15    4341       198 USA          8
```

Exploratory Data Analysis

Descriptive statistics are a good place to start when exploring data. We'll do so with the describe() function from psych:

```
describe(mpg)
#>             vars   n    mean      sd median trimmed    mad  min
#> mpg            1 392   23.45    7.81  22.75   22.99   8.60    9
#> weight         2 392 2977.58  849.40 2803.50 2916.94 948.12 1613
#> horsepower     3 392  104.47   38.49  93.50   99.82  28.91   46
#> origin*        4 392    2.42    0.81   3.00    2.53   0.00    1
#> cylinders      5 392    5.47    1.71   4.00    5.35   0.00    3
#>              max  range  skew kurtosis    se
#> mpg         46.6   37.6  0.45    -0.54  0.39
#> weight    5140.0 3527.0  0.52    -0.83 42.90
#> horsepower 230.0  184.0  1.08     0.65  1.94
#> origin*      3.0    2.0 -0.91    -0.86  0.04
#> cylinders    8.0    5.0  0.50    -1.40  0.09
```

Because *origin* is a categorical variable, we should be careful to interpret its descriptive statistics. (In fact, psych uses * to signal this warning.) We are, however, safe to analyze its one-way frequency table, which we'll do using a new dplyr function, count():

```
mpg %>%
  count(origin)
#> # A tibble: 3 x 2
#>   origin     n
#>   <chr>  <int>
#> 1 Asia      79
#> 2 Europe    68
#> 3 USA      245
```

We learn from the resulting count column n that while the majority of observations are American cars, the observations of Asian and European cars are still likely to be representative samples of their subpopulations.

Let's further break these counts down by cylinders to derive a two-way frequency table. I will combine count() with pivot_wider() to display cylinders along the columns:

```
mpg %>%
  count(origin, cylinders) %>%
  pivot_wider(values_from = n, names_from = cylinders)
#> # A tibble: 3 x 6
```

```
#>   origin    `3`    `4`    `6`    `5`    `8`
#>   <chr>   <int> <int> <int> <int> <int>
#> 1 Asia        4     69     6    NA    NA
#> 2 Europe     NA     61     4     3    NA
#> 3 USA        NA     69    73    NA   103
```

Remember that NA indicates a missing value in R, in this case because no observations were found for some of these cross-sections.

Not many cars have three- or five-cylinder engines, and *only* American cars have eight cylinders. It's common when analyzing data to have *imbalanced* datasets where there is a disproportionate number of observations in some levels. Special techniques are often needed to model such data. To learn more about working with imbalanced data, check out *Practical Statistics for Data Scientists*, 2nd edition by Peter Bruce et al. (O'Reilly).

We can also find the descriptive statistics for each level of *origin*. First, we'll use select() to choose the variables of interest, then we can use psych's describeBy() function, setting groupBy to origin:

```
mpg %>%
  select(mpg, origin) %>%
  describeBy(group = 'origin')

#>  Descriptive statistics by group
#> origin: Asia
         vars  n  mean   sd median trimmed  mad min  max range
#> mpg       1 79 30.45 6.09   31.6   30.47 6.52  18 46.6  28.6
#> origin*   2 79  1.00 0.00    1.0    1.00 0.00   1  1.0   0.0
         skew kurtosis   se
#> mpg      0.01    -0.39 0.69
#> origin*  NaN      NaN 0.00


#> origin: Europe
         vars  n mean   sd median trimmed  mad  min  max range
#> mpg       1 68 27.6 6.58     26    27.1 5.78 16.2 44.3  28.1
#> origin*   2 68  1.0 0.00      1     1.0 0.00  1.0  1.0   0.0
         skew kurtosis  se
#> mpg      0.73     0.31 0.8
#> origin*  NaN      NaN 0.0


#> origin: USA
         vars   n  mean   sd median trimmed  mad min max range
#> mpg       1 245 20.03 6.44   18.5   19.37 6.67   9  39    30
#> origin*   2 245  1.00 0.00    1.0    1.00 0.00   1   1     0
         skew kurtosis   se
#> mpg      0.83     0.03 0.41
#> origin*  NaN      NaN 0.00
```

Let's learn more about the potential relationship between *origin* and *mpg*. We'll get started by visualizing the distribution of *mpg* with a histogram, which is shown in Figure 9-1:

```
ggplot(data = mpg, aes(x = mpg)) +
  geom_histogram()
#> `stat_bin()` using `bins = 30`. Pick better value with `binwidth`.
```

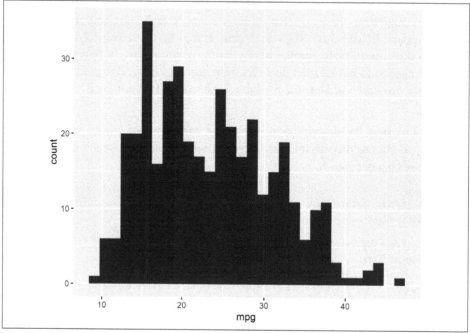

Figure 9-1. Distribution of mpg

We can now hone in on visualizing the distribution of *mpg* by *origin*. Overlaying all three levels of *origin* on one histogram could get cluttered, so a boxplot like what's shown in Figure 9-2 may be a better fit:

```
ggplot(data = mpg, aes(x = origin, y = mpg)) +
  geom_boxplot()
```

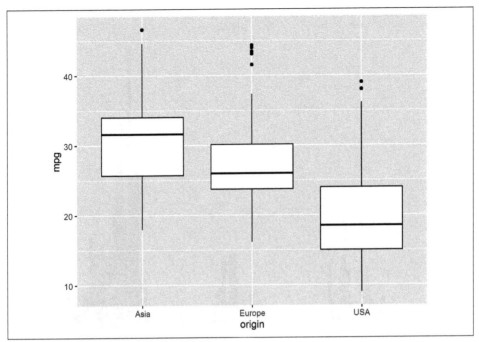

Figure 9-2. Distribution of mpg by origin

If we'd rather visualize these as histograms, and not make a mess, we can do so in R with a *facet* plot. Use `facet_wrap()` to split the ggplot2 plot into subplots, or *facets*. We'll start with a ~, or tilde operator, followed by the variable name. When you see the tilde used in R, think of it as the word "by." For example, here we are faceting a histogram by `origin`, which results in the histograms shown in Figure 9-3:

```
# Histogram of mpg, facted by origin
ggplot(data = mpg, aes(x = mpg)) +
  geom_histogram() +
  facet_grid(~ origin)
#> `stat_bin()` using `bins = 30`. Pick better value with `binwidth`.
```

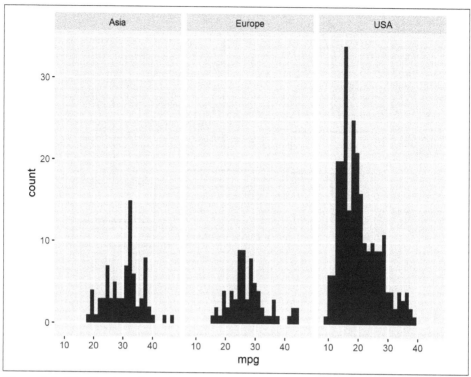

Figure 9-3. Distribution of mpg by origin

Hypothesis Testing

You could continue to explore the data using these methods, but let's move into hypothesis testing. In particular, I would like to know whether there is a significant difference in mileage between American and European cars. Let's create a new data frame containing just these observations; we'll use it to conduct a t-test.

```
mpg_filtered <- filter(mpg, origin=='USA' | origin=='Europe')
```

Testing Relationships Across Multiple Groups

We could indeed use hypothesis testing to look for a difference in mileage across American, European, and Asian cars; this is a different statistical test called *analysis of variance*, or ANOVA. It's worth exploring next on your analytics journey.

Independent Samples t-test

R includes a `t.test()` function out of the box: we need to specify where our data comes from with the `data` argument, and we'll also need to specify what *formula* to test. To do that, we'll set the relationship between independent and dependent variables with the ~ operator. The dependent variable comes in front of the ~, with independent variables following. Again, you interpret this notation as analyzing the effect of `mpg` "by" `origin`.

```
# Dependent variable ~ ("by") independent variable
t.test(mpg ~ origin, data = mpg_filtered)
#>  Welch Two Sample t-test
#>
#>     data:  mpg by origin
#>     t = 8.4311, df = 105.32, p-value = 1.93e-13
#>     alternative hypothesis: true difference in means is not equal to 0
#>     95 percent confidence interval:
#>     5.789361 9.349583
#>     sample estimates:
#>     mean in group Europe    mean in group USA
#>                 27.60294             20.03347
```

Isn't it great that R even explicitly states what our alternative hypothesis is, *and* includes the confidence interval along with the p-value? (You can tell this program was built for statistical analysis.) Based on the p-value, we will reject the null; there does appear to be evidence of a difference in means.

Let's now turn our attention to relationships between continuous variables. First, we'll use the `cor()` function from base R to print a correlation matrix. We'll do this only for the continuous variables in *mpg*:

```
select(mpg, mpg:horsepower) %>%
  cor()
#>              mpg      weight horsepower
#> mpg        1.0000000 -0.8322442 -0.7784268
#> weight    -0.8322442  1.0000000  0.8645377
#> horsepower -0.7784268  0.8645377  1.0000000
```

We can use `ggplot2` to visualize, for example, the relationship between weight and mileage, as in Figure 9-4:

```
ggplot(data = mpg, aes(x = weight,y = mpg)) +
  geom_point() + xlab('weight (pounds)') +
  ylab('mileage (mpg)') + ggtitle('Relationship between weight and mileage')
```

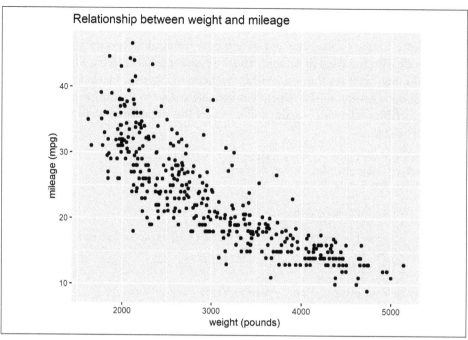

Figure 9-4. Scatterplot of weight by mpg

Alternatively, we could use the `pairs()` function from base R to produce a pairplot of all combinations of variables, laid out similarly to a correlation matrix. Figure 9-5 is a pairplot of selected variables from *mpg*:

```
select(mpg, mpg:horsepower) %>%
  pairs()
```

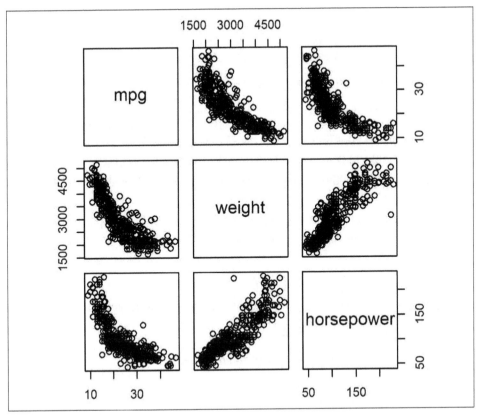

Figure 9-5. Pairplot

Linear Regression

We're ready now for linear regression, using base R's lm() function (this is short for *linear model*). Similar to t.test(), we will specify a dataset and a formula. Linear regression returns a fair amount more output than a t-test, so it's common to assign the results to a new object in R first, then explore its various elements separately. In particular, the summary() function provides a helpful overview of the regression model:

```
mpg_regression <- lm(mpg ~ weight, data = mpg)
summary(mpg_regression)

#>  Call:
#>  lm(formula = mpg ~ weight, data = mpg)
#>
#>  Residuals:
#>      Min      1Q   Median      3Q      Max
#>  -11.9736  -2.7556  -0.3358  2.1379  16.5194
#>
```

```
#>    Coefficients:
#>                 Estimate Std. Error t value Pr(>|t|)
#>    (Intercept) 46.216524   0.798673   57.87   <2e-16 ***
#>    weight      -0.007647   0.000258  -29.64   <2e-16 ***
#>    ---
#>    Signif. codes:  0 '***' 0.001 '**' 0.01 '*' 0.05 '.' 0.1 ' ' 1
#>
#>    Residual standard error: 4.333 on 390 degrees of freedom
#>    Multiple R-squared:  0.6926,     Adjusted R-squared:  0.6918
#>    F-statistic: 878.8 on 1 and 390 DF,  p-value: < 2.2e-16
```

This output should look familiar. Here you'll see the coefficients, p-values, and R-squared, among other figures. Again, there does appear to be a significant influence of weight on mileage.

Last but not least, we can fit this regression line over the scatterplot by including `geom_smooth()` in our `ggplot()` function, setting `method` to `lm`. This results in Figure 9-6:

```
ggplot(data = mpg, aes(x = weight, y = mpg)) +
  geom_point() + xlab('weight (pounds)') +
  ylab('mileage (mpg)') + ggtitle('Relationship between weight and mileage') +
  geom_smooth(method = lm)
#> `geom_smooth()` using formula 'y ~ x'
```

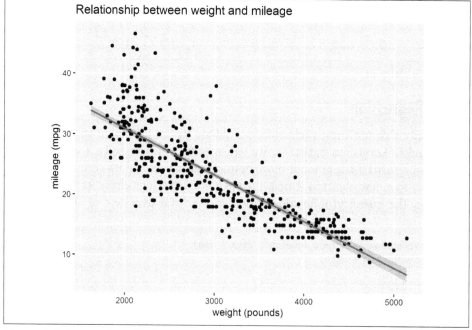

Figure 9-6. Scatterplot with fit regression line of weight by mpg

Train/Test Split and Validation

Chapter 5 briefly reviewed how machine learning relates to working with data more broadly. A technique popularized by machine learning that you may encounter in your data analytics work is the *train/test split*. The idea here is to *train* the model on a subset of your data, then *test* it on another subset. This provides assurance that the model doesn't just work on one particular sampling of observations, but can generalize to the wider population. Data scientists are often especially interested in how well the model does at making predictions on the testing data.

Let's split our *mpg* dataset in R, train the linear regression model on part of the data, and then test it on the remainder. To do so, we'll use the `tidymodels` package. While not part of the `tidyverse`, this package is built along the same principles and thus works well with it.

You may remember in Chapter 2 that, because we were using random numbers, the results you saw in your workbook were different than what was documented in the book. Because we'll again be splitting our dataset randomly here, we could encounter that same problem. To avoid that, we can set the *seed* of R's random number generator, which results in the same series of random numbers being generated each time. This can be done with the `set.seed()` function. You can set it to any number; 1234 is common:

```
set.seed(1234)
```

To begin the split, we can use the aptly named `initial_split()` function; from there, we'll subset our data into training and testing datasets with the `training()` and `testing()` functions, respectively.

```
mpg_split <- initial_split(mpg)
mpg_train <- training(mpg_split)
mpg_test <- testing(mpg_split)
```

By default, `tidymodels` splits the data's observations into two groups at random: 75% of the observations went to the training group, the remainder to the test. We can confirm that with the `dim()` function from base R to get the number of rows and columns in each dataset, respectively:

```
dim(mpg_train)
#> [1] 294    5
dim(mpg_test)
#> [1] 98  5
```

At 294 and 98 observations, our training and testing sample sizes should be sufficiently large for reflective statistical inference. While it's not often a consideration for the massive datasets used in machine learning, adequate sample size can be a limitation when splitting data.

It's possible to split the data into other proportions than 75/25, to use special techniques for splitting the data, and so forth. For more information, check the tidymodels documentation; until you become more comfortable with regression analysis, the defaults are fine.

To build our training model, we'll first *specify* what type of model it is with the linear_reg() function, then *fit* it. The inputs of the fit() function should look familiar to you, except this time we are using the training subset of *mpg* only.

```
# Specify what kind of model this is
lm_spec <- linear_reg()

# Fit the model to the data
lm_fit <- lm_spec %>%
  fit(mpg ~ weight, data = mpg_train)
#> Warning message:
#> Engine set to `lm`.
```

You will see from your console output that the lm() function from base R, which you've used before, was used as the *engine* to fit the model.

We can get the coefficients and p-values of our training model with the tidy() function, and its performance metrics (such as R-squared) with glance().

```
tidy(lm_fit)
#> # A tibble: 2 x 5
#>   term         estimate std.error statistic   p.value
#>   <chr>           <dbl>     <dbl>     <dbl>     <dbl>
#> 1 (Intercept) 47.3       0.894       52.9 1.37e-151
#> 2 weight      -0.00795   0.000290   -27.5 6.84e- 83
#>
glance(lm_fit)
#> # A tibble: 1 x 12
#>   r.squared adj.r.squared sigma statistic  p.value    df logLik   AIC
#>       <dbl>         <dbl> <dbl>     <dbl>    <dbl> <dbl>  <dbl> <dbl>
#> 1     0.721         0.720  4.23      754. 6.84e-83     1  -840. 1687.
#> # ... with 4 more variables: BIC <dbl>, deviance <dbl>,
#> #   df.residual <int>, nobs <int>
```

This is great, but what we *really* want to know is how well this model performs when we apply it to a new dataset; this is where the test split comes in. To make predictions on `mpg_test`, we'll use the `predict()` function. I will also use `bind_cols()` to add the column of predicted Y-values to the data frame. This column by default will be called `.pred`.

```
mpg_results <- predict(lm_fit, new_data = mpg_test) %>%
  bind_cols(mpg_test)

mpg_results
#> # A tibble: 98 x 6
#>     .pred   mpg weight horsepower origin cylinders
#>     <dbl> <dbl>  <dbl>      <dbl> <chr>      <dbl>
#>  1   20.0    16   3433        150 USA            8
#>  2   16.7    15   3850        190 USA            8
#>  3   25.2    18   2774         97 USA            6
#>  4   30.3    27   2130         88 Asia           4
#>  5   28.0    24   2430         90 Europe         4
#>  6   21.0    19   3302         88 USA            6
#>  7   14.2    14   4154        153 USA            8
#>  8   14.7    14   4096        150 USA            8
#>  9   29.6    23   2220         86 USA            4
#> 10   29.2    24   2278         95 Asia           4
#> # ... with 88 more rows
```

Now that we've applied the model to this new data, let's evaluate its performance. We can, for example, find its R-squared with the `rsq()` function. From our `mpg_results` data frame, we'll need to specify which column contains the actual Y values with the `truth` argument, and which are predictions with the `estimate` column.

```
rsq(data = mpg_results, truth = mpg, estimate = .pred)
#> # A tibble: 1 x 3
#>   .metric .estimator .estimate
#>   <chr>   <chr>          <dbl>
#> 1 rsq     standard       0.606
```

At an R-squared of 60.6%, the model derived from the training dataset explains a fair amount of variability in the testing data.

Another common evaluation metric is the root mean square error (RMSE). You learned about the concept of *residuals* in Chapter 4 as the difference between actual and predicted values; RMSE is the standard deviation of the residuals and thus an estimate of how spread errors tend to be. The `rmse()` function returns the RMSE.

```
rmse(data = mpg_results, truth = mpg, estimate = .pred)
#> # A tibble: 1 x 3
#>   .metric .estimator .estimate
#>   <chr>   <chr>          <dbl>
#> 1 rmse    standard        4.65
```

Because it's relative to the scale of the dependent variable, there's no one-size-fits-all way to evaluate RMSE, but between two competing models using the same data, a smaller RMSE is preferred.

`tidymodels` makes numerous techniques available for fitting and evaluating models in R. We've looked at a regression model, which takes a continuous dependent variable, but it's also possible to build *classification* models, where the dependent variable is categorical. This package is a relative newcomer to R, so there is somewhat less literature available, but expect more to come as the package grows in popularity.

Conclusion

There is, of course, much more you could do to explore and test the relationships in this and other datasets, but the steps we've taken here serve as a solid opening. Earlier, you were able to conduct and interpret this work in Excel, and now you've leaped into doing it in R.

Exercises

Take a moment to try your hand at analyzing a familiar dataset with familiar steps, now using R. At the end of Chapter 4, you practiced analyzing data from the `ais` dataset in the book repository (*https://oreil.ly/egOx1*). This data is available in the R package `DAAG`; try installing and loading it from there (it is available as the object `ais`). Do the following:

1. Visualize the distribution of red blood cell count (*rcc*) by sex (*sex*).
2. Is there a significant difference in red blood cell count between the two groups of sex?
3. Produce a correlation matrix of the relevant variables in this dataset.
4. Visualize the relationship of height (*ht*) and weight (*wt*).
5. Regress *ht* on *wt*. Find the equation of the fit regression line. Is there a significant relationship? What percentage of the variance in *ht* is explained by *wt*?
6. Split your regression model into training and testing subsets. What is the R-squared and RMSE on your test model?

PART III

From Excel to Python

## CHAPTER 10

First Steps with Python for Excel Users

Created in 1991 by Guido van Rossum, Python is a programming language that, like R, is free and open source. At the time, van Rossum was reading the scripts from *Monty Python's Flying Circus* and decided to name the language after the British comedy. Unlike R, which was designed explicitly for data analysis, Python was developed as a general-purpose language meant to do things like interact with operating systems, handle processing errors, and so forth. This has some important implications for how Python "thinks" and works with data. For example, you saw in Chapter 7 that R has a built-in tabular data structure. This isn't the case in Python; we'll need to rely more heavily on external packages to work with data.

That's not necessarily a problem: Python, like R, has thousands of packages maintained by a thriving contributor community. You'll find Python used for everything from mobile app development to embedded devices to, yes, data analytics. Its diverse user base is growing rapidly, and Python has become one of the most popular programming languages not just for analytics but for computing generally.

 Python was conceived as a general-purpose programming language, while R was bred specifically with statistical analysis in mind.

Downloading Python

The Python Software Foundation (*https://python.org*) maintains the "official" Python source code. Because Python is open source, anyone is available to take, add to, and redistribute Python code. Anaconda is one such Python distribution and is the suggested installation for this book. It's maintained by a for-profit company of the same name and is available in paid tiers; we'll be using the free Individual Edition. Python is now on its third version, Python 3. You can download the latest release of Python 3 at Anaconda's website (*https://oreil.ly/3RYeQ*).

Python 2 and Python 3

Python 3, released in 2008, made significant changes to the language and importantly was not backward compatible with code from Python 2. This means that code written for Python 2 may not necessarily run on Python 3, and vice versa. At the time of writing, Python 2 has been officially retired, although you may encounter some references and code remnants in your Python journey.

In addition to a simplified installation of Python, Anaconda comes with extras, including some popular packages which we'll use later in the book. It also ships with a web application that we'll use to work with Python: the Jupyter Notebook.

Getting Started with Jupyter

As mentioned in Chapter 6, R was modeled after the S program for EDA. Because of the iterative nature of EDA, the expected workflow of the language is to execute and explore the output of selected lines of code. This makes it easy to conduct data analysis directly from an R script, *.r*. We used the RStudio IDE to provide additional support for our programming session, such as dedicated panes for help documentation and information about the objects in our environment.

By contrast, Python in some ways behaves more like "lower-level" programming languages, where code needs first to be compiled into a machine-readable file, and *then* run. This can make it relatively trickier to conduct piecemeal data analysis from a Python script, *.py*. This pain point of working with Python for statistical and, more broadly, scientific computing caught the attention of physicist and software developer Fernando Pérez, who with colleagues in 2001 launched the IPython project to make a more interactive interpreter for Python (IPython as a playful shorthand for "interactive Python"). One result of this initiative was a new type of file, the *IPython Notebook*, denoted with the *.ipynb* file extension.

This project gained traction and in 2014, IPython was spun into the broader Project Jupyter, a language-agnostic initiative to develop interactive, open source computing software. Thus, the IPython Notebook became the Jupyter Notebook while retaining the *.ipynb* extension. Jupyter notebooks run as interactive web applications that allow users to combine code with text, equations, and more to create media-rich interactive documents. In fact, Jupyter was named in part as an homage to the notebooks Galileo used to record his discovery of the planet Jupiter's moons. A *kernel* is used behind the scenes to execute the notebook's code. By downloading Anaconda, you've set up all these necessary parts to execute Python from a Jupyter notebook: now you just need to launch a session.

RStudio, Jupyter Notebooks, and Other Ways to Code

You may be unhappy to leave RStudio to learn yet another interface. But remember that code and application are often decoupled in open source frameworks; it's easy to "remix" these languages and platforms. For example, R is one of the many dozens of languages with a kernel for Jupyter. Along with the Galileo reference, Jupyter is *also* a portmanteau of its three core supported languages: Julia, Python, and R.

Conversely, it's possible to execute Python scripts from inside RStudio with the help of R's `reticulate` package, which can more broadly be used to run Python code from R. This means it's possible to, for example, import and manipulate data in Python and then use R to visualize the results. Other popular programs for working with Python code include PyCharm and Visual Studio Code. RStudio also has its own notebook application with R Notebooks. The same concept as Jupyter of interspersing code and text applies, and it supports several languages including R and Python.

As you're starting to see, there's a whole galaxy of tools available for coding in R and Python, more than can be covered in this book. Our focus has been on R scripts from RStudio and Python from Jupyter Notebooks because they are both relatively more beginner-friendly and common than other configurations. Once you get comfortable with these workflows, search online for the development environments mentioned here. As you continue to learn, you'll pick up even more ways to interact with these languages.

The steps for launching a Jupyter notebook vary for Windows and Mac computers. On Windows, open the Start menu, then search for and launch `Anaconda Prompt`. This is a command-line tool for working with your Anaconda distribution and yet another way to interact with Python code. For a further introduction to running Python from the command line with the experience of an Excel user in mind, check out Felix Zumstein's *Python for Excel* (O'Reilly). From inside the Anaconda prompt, enter `jupyter notebook` at the cursor and hit `Enter`. Your command will resemble the following, but with a different home directory path:

```
(base) C:\Users\User> jupyter notebook
```

On a Mac, open Launchpad, then search for and launch Terminal. This is the command-line interface that ships with Macs and can be used to communicate with Python. From inside the Terminal prompt, enter `jupyter notebook` at the cursor and hit `Enter`. Your command line will resemble the following, but with a different home directory path:

```
user@MacBook-Pro ~ % jupyter notebook
```

After doing this on either system, a couple of things will happen: first, an additional terminal-like window will launch on your computer. *Do not close this window.* This is what establishes the connection to the kernel. Additionally, the Jupyter notebook interface should automatically open in your default web browser. If it does not, the terminal-like window will include a link that you can paste into your browser. Figure 10-1 shows what you should see in your browser. Jupyter launches with a File-Explorer-like interface. You can now navigate to the folder in which you'd like to save your notebooks.

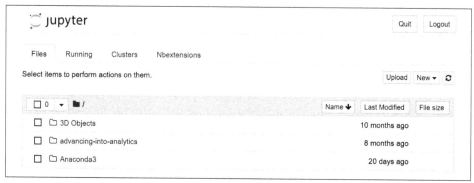

Figure 10-1. Jupyter landing page

To open a new notebook, head to the upper-right side of your browser window and select New → Notebook → Python 3. A new tab will open with a blank Jupyter notebook. Like RStudio, Jupyter provides far more features than we can cover in an introduction; we'll focus on the key pieces to get you started. The four main components of a Jupyter notebook are labeled in Figure 10-2; let's walk through each.

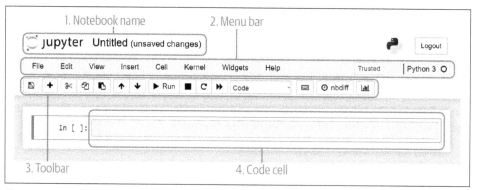

Figure 10-2. Elements of the Jupyter interface

First, the notebook name: this is the name of our *.ipynb* file. You can rename the notebook by clicking and writing over the current name.

Next, the menu bar. This contains different operations for working with your notebook. For example, under File you can open and close notebooks. Saving them isn't much of an issue, because Jupyter notebooks are autosaved every two minutes. If you ever need to convert your notebook to a *.py* Python script or other common file type, you can do so by going to File → Download as. There's also a Help section containing several guides and links to reference documentation. You can learn about Jupyter's many keyboard shortcuts from this menu.

Earlier, I mentioned that the *kernel* is how Jupyter interacts with Python under the hood. The *Kernel* option in the menu bar contains helpful operations for working with it. Computers being what they are, sometimes all that's needed to get your Python code working is to restart the kernel. You can do this by going to Kernel → Restart.

Immediately underneath the menu bar is the toolbar. This contains helpful icons for working with your notebook, which can be more convenient than navigating through the menu: for example, several icons here relate to interacting with the kernel.

You can also insert and relocate *cells* in your notebook, where you'll be spending most of your time in Jupyter. To get started, let's do one last thing with the toolbar: you'll find a drop-down menu there currently set to Code; change it to Markdown.

Now, navigate to your first code cell and type in the phrase, `Hello, Jupyter!` Head back to the toolbar and select the Run icon. A couple of things will happen. First, you'll see that your `Hello, Jupyter!` cell will render to look as it might in a word processing document. Next, you'll see that a new code cell is placed underneath your previous one, and that it's set for you to enter more information. Your notebook should resemble Figure 10-3.

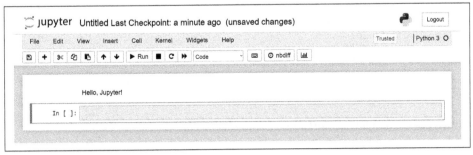

Figure 10-3. "Hello, Jupyter!"

Now, go back to the toolbar and again choose "Markdown" from the drop-down menu. As you're beginning to find out, Jupyter notebooks consist of modular cells that can be of different types. We'll focus on the two most common: Markdown and Code. Markdown is a plain-text markup language that uses regular keyboard characters to format text.

Insert the following text into your blank cell:

```
# Big Header 1
## Smaller Header 2
### Even smaller headers
#### Still more

*Using one asterisk renders italics*

**Using two asterisks renders bold**

- Use dashes to...
- Make bullet lists

Refer to code without running it as `fixed-width text`
```

Now run the cell: you can do this either from the toolbar or with the shortcut Alt + Enter for Windows, Option + Return for Mac. Your selection will render as in Figure 10-4.

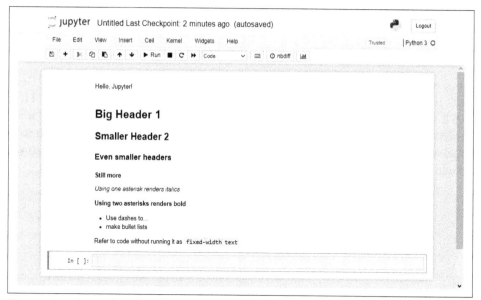

Figure 10-4. Examples of Markdown formatting in Jupyter

To learn more about Markdown, return to the Help section of the menu bar. It's worth studying up to build elegant notebooks, which can include images, equations, and more. But in this book, we'll focus on the *code* block, as that's where executable Python goes. You should now be on your third code cell; you can leave this one as a Code format. Finally, we'll get coding in Python.

Python can be used as a fancy calculator, just like Excel and R. Table 10-1 lists some common arithmetic operators in Python.

Table 10-1. Common arithmetic operators in Python

Operator	Description
+	Addition
-	Subtraction
*	Multiplication
/	Division
**	Exponent
%%	Modulo
//	Floor division

Enter in the following arithmetic, then run the cells:

```
In [1]: 1 + 1
Out[1]: 2

In [2]: 2 * 4
Out[2]: 8
```

As Jupyter code blocks are executed, they are given numbered labels of their inputs and outputs with In [] and Out [], respectively.

Python also follows the order of operations; let's try running a few examples from within the same cell:

```
In [3]: # Multiplication before addition
        3 * 5 + 6
        2 / 2 - 7 # Division before subtraction
Out[3]: -6.0
```

By default, Jupyter notebooks will only return the output of the last-run code within a cell, so we'll break this into two. You can split a cell at the cursor on either Windows or Mac with the keyboard shortcut Ctrl + Shift + - (Minus):

```
In [4]:  # Multiplication before addition
         3 * 5 + 6

Out[4]: 21

In [5]:  2 / 2 - 7 # Division before subtraction

Out[5]: -6.0
```

And yes, Python also uses code comments. Similar to R, they start with a hash, and it's also preferable to keep them to separate lines.

Like Excel and R, Python includes many functions for both numbers and characters:

```
In [6]: abs(-100)

Out[6]: 100
```

```
In [7]: len('Hello, world!')

Out[7]: 13
```

Unlike Excel, but like R, Python is case-sensitive. That means *only* abs() works, not ABS() or Abs().

```
In [8]:  ABS(-100)
```

```
--------------------------------------------------------------------
NameError                             Traceback (most recent call last)
<ipython-input-20-a0f3f8a69d46> in <module>
----> 1 print(ABS(-100))
      2 print(Abs(-100))

NameError: name 'ABS' is not defined
```

Python and Indentation

In Python, whitespace is more than a suggestion: it can be a *necessity* for code to run. That's because the language relies on proper indentation to compile and execute code blocks, or pieces of Python that are meant to be executed as a unit. You won't run into the problem in this book, but as you continue to experiment with other features of Python, such as how to write functions or loops, you'll see how prevalent and critical indentation is to the language.

Similar to R, you can use the ? operator to get information about functions, packages, and more. A window will open as in Figure 10-5, which you can then expand or open in a new window.

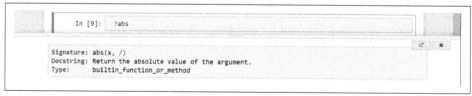

Figure 10-5. Launching documentation in Jupyter notebooks

Comparison operators mostly work the same in Python as in R and Excel; in Python, results are either returned as True or False.

```
In [10]: # Is 3 greater than 4?
         3 > 4
```

```
Out[10]: False
```

As with R, you check for whether two values are equal with ==; a single equals sign = is used to assign objects. We'll stick with = throughout to assign objects in Python.

```
In [11]:  # Assigning an object in Python
          my_first_object = abs(-100)
```

You may have noticed there was no Out [ ] component of cell 11. That's because we only *assigned* the object; we didn't print anything. Let's do that now:

```
In [12]: my_first_object
```

```
Out[12]: 100
```

Object names in Python must start with a letter or underscore, and the rest of the name can contain only letters, numbers, or underscores. There are also a few off-limit keywords. Again, you're left with broad license for naming objects in Python, but just because you *can* name an object scooby_doo doesn't mean you should.

Python and PEP 8

The Python Foundation uses Python Enhancement Proposals (PEPs) to announce changes or new features to the language. PEP 8 offers a style guide that is the universal standard for writing Python code. Among its many rules and guidelines are conventions for naming objects, adding comments, and more. You can read the full PEP 8 guide on the Python Foundation's website (*https://oreil.ly/KdmIf*).

Just like in R, our objects in Python can consist of different data types. Table 10-2 shows some basic Python types. Do you see the similarities and differences to R?

Table 10-2. Basic object types in Python

Data type	Example
String	'Python', 'G. Mount', 'Hello, world!'
Float	6.2, 4.13, 3.1
Integer	3, -1, 12
Boolean	True, False

Let's assign some objects. We can find what type they are with the type() function:

```
In [13]:  my_string = 'Hello, world'
          type(my_string)
```

```
Out[13]: str
```

```
In [14]: # Double quotes work for strings, too
         my_other_string = "We're able to code Python!"
         type(my_other_string)
```

```
Out[14]: str
```

```
In [15]: my_float = 6.2
         type(my_float)

Out[15]: float


In [16]: my_integer = 3
         type(my_integer)

Out[16]: int

In [17]: my_bool = True
         type(my_bool)

Out[17]: bool
```

You've worked with objects in R, so you're probably not surprised that it's possible to use them as part of Python operations.

```
In [18]: # Is my_float equal to 6.1?
         my_float == 6.1

Out[18]: False

In [19]: # How many characters are in my_string?
         # (Same function as Excel)
         len(my_string)

Out[19]: 12
```

Closely related to functions in Python are *methods*. A method is affixed to an object with a period and does something to that object. For example, to capitalize all letters in a string object, we can use the upper() method:

```
In [20]: my_string.upper()

Out[20]: 'HELLO, WORLD'
```

Functions and methods are both used to perform operations on objects, and we'll use both in this book. As you are probably hoping, Python, like R, can store multiple values in a single object. But before getting into that, let's consider how *modules* work in Python.

Modules in Python

Python was designed as a general-purpose programming language, so even some of the simplest functions for working with data aren't available out of the box. For example, we won't have luck finding a function to take the square root of a number:

```
In [21]:  sqrt(25)
```

```
----------------------------------------------------------
NameError                        Traceback (most recent call last)
<ipython-input-18-1bf613b64533> in <module>
----> 1 sqrt(25)

NameError: name 'sqrt' is not defined
```

This function *does* exist in Python. But to access it, we'll need to bring in a *module*, which is like a package in R. Several modules come installed with Python as part of the Python Standard Library; for example, the math module contains many mathematical functions, including sqrt(). We can call this module into our session with the import statement:

```
In [22]:  import math
```

Statements are instructions telling the interpreter what to do. We just told Python to, well, *import* the math module. The sqrt() function should now be available to us; give it a try:

```
In [23]:  sqrt(25)
```

```
----------------------------------------------------------
NameError                        Traceback (most recent call last)
<ipython-input-18-1bf613b64533> in <module>
----> 1 sqrt(25)

NameError: name 'sqrt' is not defined
```

Honestly, I'm not fibbing about a sqrt() function. The reason we're still getting errors is we need to explicitly tell Python *where* that function comes from. We can do that by prefixing the module name before the function, like so:

```
In [24]:  math.sqrt(25)
```

```
Out[24]: 5.0
```

The Standard Library is full of helpful modules. Then there are the thousands of third-party modules, bundled into *packages* and submitted to the Python Package Index. pip is the standard package-management system; it can be used to install from the Python Package Index as well as outside sources.

The Anaconda distribution has done much of the lifting for working with packages. First off, some of the most popular Python packages come preinstalled. Additionally, Anaconda includes features to ensure all packages on your machine are compatible. For this reason, it's preferred to install packages directly from Anaconda rather than from pip. Python package installation is generally done from the command line; you worked there earlier when you were in the Anaconda Prompt (Windows) or Terminal (Mac). However, we can execute command-line code from Jupyter by including an exclamation mark in front of it. Let's install `plotly`, a popular package for data visualization, from Anaconda. The statement to use is `conda install`:

```
In [25]: !conda install plotly
```

Not all packages are available to download from Anaconda; in that case, we can install via `pip`. Let's do it for the `pyxlsb` package, which can be used to read binary `.xlsb` Excel files into Python:

```
In [26]: !pip install pyxlsb
```

Although downloading packages right from Jupyter is convenient, it can be an unpleasant surprise for others to try running your notebook only to get hit with lengthy or unnecessary downloads. That's why it's common to comment out install commands, a convention I follow in the book repository.

 If you're using Anaconda to run Python, it's best to install things first via `conda` and only then install via `pip` if the package is not available.

Upgrading Python, Anaconda, and Python packages

Table 10-3 lists several other helpful commands for maintaining your Python environment. You can also install and maintain Anaconda packages from a point-and-click interface using the Anaconda Navigator, which is installed with Anaconda Individual Edition. To get started, launch the application on your computer, then navigate to the Help menu to read the documentation for more.

Table 10-3. Helpful commands for maintaining Python packages

Command	Description
`conda update anaconda`	Updates Anaconda distribution
`conda update python`	Updates Python version
`conda update -- all`	Updates all possible packages downloaded via `conda`
`pip list -- outdated`	Lists all packages downloaded via `pip` that can be updated

Conclusion

In this chapter, you learned how to work with objects and packages in Python and got the hang of working with Jupyter notebooks.

Exercises

The following exercises provide additional practice and insight on these topics:

1. From a new Jupyter notebook, do the following:
   - Assign the sum of 1 and –4 as a.
   - Assign the absolute value of a as b.
   - Assign b minus 1 as d.
   - Is d greater than 2?

2. The Python Standard Library includes a module random containing a function randint(). This function works like RANDBETWEEN() in Excel; for example, randint(1, 6) will return an integer between 1 and 6. Use this function to find a random number between 0 and 36.

3. The Python Standard Library also includes a module called this. What happens when you import that module?

4. Download the xlutils package from Anaconda, then use the ? operator to retrieve the available documentation.

I will again encourage you to begin using the language as soon as possible in your everyday work, even if at first it's just as a calculator. You can also try performing the same tasks in R *and* Python, then comparing and contrasting. If you learned by relating R to Excel, the same will work for relating Python to R.

## CHAPTER 11

Data Structures in Python

In Chapter 10, you learned about simple Python object types like strings, integers, and Booleans. Now let's look at grouping multiple values together in what's called a *collection*. Python by default comes with several collection object types. We'll start this chapter with the *list*. We can put values into a list by separating each entry with commas and placing the results inside square brackets:

```
In [1]: my_list = [4, 1, 5, 2]
        my_list

Out[1]: [4, 1, 5, 2]
```

This object contains all integers, but itself is *not* an integer data type: it is a *list*.

```
In [2]: type(my_list)

Out[2]: list
```

In fact, we can include all different sorts of data inside a list…even other lists.

```
In [3]: my_nested_list = [1, 2, 3, ['Boo!', True]]
        type(my_nested_list)

Out[3]: list
```

Other Collection Types in Base Python

Python includes several other built-in collection object types besides the list, most notably the *dictionary*, along with still more in the Standard Library's collections module. Collection types vary by how they store values and can be indexed or modified.

As you're seeing, lists are quite versatile for storing data. But right now, we're really interested in working with something that could function like an Excel range or R vector, and then move into tabular data. Does a simple list fit the bill? Let's give it a whirl by trying to multiply my_list by two.

```
In [4]:  my_list * 2
```

```
Out[4]: [4, 1, 5, 2, 4, 1, 5, 2]
```

This is probably *not* what you are looking for: Python took you literally and, well, doubled your *list*, rather than the *numbers inside* your list. There are ways to get what we want here on our own: if you've worked with loops before, you could set one up here to multiply each element by two. If you've not worked with loops, that's fine too: the better option is to import a module that makes it easier to perform computations in Python. For that, we'll use numpy, which is included with Anaconda.

NumPy arrays

```
In [5]:  import numpy
```

As its name suggests, numpy is a module for numerical computing in Python and has been foundational to Python's popularity as an analytics tool. To learn more about numpy, visit the Help section of Jupyter's menu bar and select "NumPy reference." We'll focus for right now on the numpy *array*. This is a collection of data with all items of the same type and that can store data in up to any number, or *n* dimensions. We'll focus on a *one-dimensional* array and convert our first one from a list using the array() function:

```
In [6]:  my_array = numpy.array([4, 1, 5, 2])
         my_array
```

```
Out[6]: array([4, 1, 5, 2])
```

At first glance a numpy array looks a *lot* like a list; after all, we even created this one *from* a list. But we can see that it really is a different data type:

```
In [7]: type(my_list)
```

```
Out[7]: list
```

```
In [8]: type(my_array)
```

```
Out[8]: numpy.ndarray
```

Specifically, it's an ndarray, or *n*-dimensional array. Because it's a different data structure, it may behave differently with operations. For example, what happens when we multiply a numpy array?

```
In [9]: my_list * 2

Out[9]: [4, 1, 5, 2, 4, 1, 5, 2]


In [10]: my_array * 2

Out[10]: array([ 8,  2, 10,  4])
```

In many ways this behavior should remind you of an Excel range or an R vector. And indeed, like R vectors, numpy arrays will *coerce* data to be of the same type:

```
In [11]: my_coerced_array = numpy.array([1, 2, 3, 'Boo!'])
         my_coerced_array

Out[11]: array(['1', '2', '3', 'Boo!'], dtype='<U11')
```

Data Types in NumPy and Pandas

You'll notice that data types in numpy and later pandas work a bit differently than standard Python. These so-called dtypes are built to read and write data quickly and work with low-level programming languages like C or Fortran. Don't worry too much about the specific dtypes being used; focus on the general kind of data you're working with, such as floating point, string, or Boolean.

As you're seeing, numpy is a lifesaver for working with data in Python. Plan to import it *a lot*…which means typing it a lot. Fortunately, you can lighten the load with *aliasing*. We'll use the **as** keyword to give numpy its conventional alias, np:

```
In [12]: import numpy as np
```

This gives the module a temporary, more manageable name. Now, each time we want to call in code from numpy during our Python session, we can refer to its alias.

```
In [13]: import numpy as np
         # numpy also has a sqrt() function:
         np.sqrt(my_array)

Out[13]: array([2.        , 1.        , 2.23606798, 1.41421356])
```

 Remember that aliases are *temporary* to your Python session. If you restart your kernel or start a new notebook, the alias won't work anymore.

Indexing and Subsetting NumPy Arrays

Let's take a moment to explore how to pull individual items from a `numpy` array, which we can do by affixing its index number in square brackets directly next to the object name:

```
In [14]: # Get second element... right?
         my_array[2]

Out[14]: 5
```

For example, we just pulled the second element from our array...*or did we?* Let's revisit `my_array`; what is *really* showing in the second position?

```
In [15]: my_array

Out[15]: array([4, 1, 5, 2])
```

It appears that 1 is in the second position, and 5 is actually in the *third*. What explains this discrepancy? As it turns out, it's because Python counts things differently than you and I usually do.

As a warm-up to this strange concept, imagine being so excited to get your hands on a new dataset that you download it several times. That hastiness leaves you with a series of files named like this:

- *dataset.csv*
- *dataset (1).csv*
- *dataset (2).csv*
- *dataset (3).csv*

As humans, we tend to start counting things at one. But computers often start counting at *zero*. Multiple file downloads is one example: our second file is actually named `dataset (1)`, not `dataset (2)`. This is called *zero-based indexing*, and it happens *all over* in Python.

Zero- and One-Based Indexing

Computers often count from zero, but not all the time. In fact, Excel and R both implement *one-based* indexing, where the first element is considered to be in position one. Programmers can have strong opinions about which is a better design, but you should be comfortable working in both frameworks.

This is all to say that, to Python, indexing something with the number 1 returns the value in the *second* position, indexing with 2 returns the third, and so on.

```
In [16]: # *Now* let's get the second element
         my_array[1]
```

```
Out[16]: 1
```

It's also possible to subset a selection of consecutive values, called *slicing* in Python. Let's try finding the second through fourth elements. We already got the zero-based kicker out of the way; how hard could this be?

```
In [17]: # Get second through fourth elements... right?
         my_array[1:3]
```

```
Out[17]: array([1, 5])
```

But wait, there's more. In addition to being zero-indexed, slicing is *exclusive* of the ending element. That means we need to "add 1" to the second number to get our intended range:

```
In [18]: # *Now* get second through fourth elements
         my_array[1:4]
```

```
Out[18]: array([1, 5, 2])
```

There's much more you can do with slicing in Python, such as starting at the *end* of an object or selecting all elements from the start to a given position. For now, the important thing to remember is that *Python uses zero-based indexing*.

Two-dimensional numpy arrays can serve as a tabular Python data structure, but all elements must be of the same data type. This is rarely the case when we're analyzing data in a business context, so to meet this requirement we'll move to pandas.

Introducing Pandas DataFrames

Named after the *panel data* of econometrics, pandas is especially helpful for manipulating and analyzing tabular data. Like numpy, it comes installed with Anaconda. The typical alias is pd:

```
In [19]: import pandas as pd
```

The pandas module leverages numpy in its code base, and you will see some similarities between the two. pandas includes, among others, a one-dimensional data structure called a *Series*. But its most widely used structure is the two-dimensional *DataFrame* (sound familiar?). It's possible to create a DataFrame from other data types, including numpy arrays, using the DataFrame function:

```
In [20]: record_1 = np.array(['Jack', 72, False])
         record_2 = np.array(['Jill', 65, True])
         record_3 = np.array(['Billy', 68, False])
         record_4 = np.array(['Susie', 69, False])
         record_5 = np.array(['Johnny', 66, False])
```

```
roster = pd.DataFrame(data = [record_1,
    record_2, record_3, record_4, record_5],
    columns = ['name', 'height', 'injury'])

roster
```

Out[20]:

```
     name height injury
0    Jack     72  False
1    Jill     65   True
2   Billy     68  False
3   Susie     69  False
4  Johnny     66  False
```

DataFrames generally include named *labels* for each column. There will also be an *index* running down the rows, which by default starts at (you guessed it) 0. This is a pretty small dataset to explore, so let's find something else. Unfortunately, Python does not include any DataFrames out of the gate, but we can find some with the seaborn package. seaborn also comes installed with Anaconda and is often aliased as sns. The get_dataset_names() function will return a list of DataFrames available to use:

```
In [21]: import seaborn as sns
         sns.get_dataset_names()
```

Out[21]:

```
    ['anagrams', 'anscombe', 'attention', 'brain_networks', 'car_crashes',
     'diamonds', 'dots', 'exercise', 'flights', 'fmri', 'gammas',
     'geyser', 'iris', 'mpg', 'penguins', 'planets', 'tips', 'titanic']
```

Does *iris* sound familiar? We can load it into our Python session with the load_data set() function, and print the first five rows with the head() method.

```
In [22]: iris = sns.load_dataset('iris')
         iris.head()
```

Out[22]:

```
   sepal_length  sepal_width  petal_length  petal_width species
0           5.1          3.5           1.4          0.2  setosa
1           4.9          3.0           1.4          0.2  setosa
2           4.7          3.2           1.3          0.2  setosa
3           4.6          3.1           1.5          0.2  setosa
4           5.0          3.6           1.4          0.2  setosa
```

Importing Data in Python

As with R, it's most common to read in data from external files, and we'll need to deal with directories to do so. The Python Standard Library includes the os module for working with file paths and directories:

```
In [23]: import os
```

For this next part, have your notebook saved in the main folder of the book repository. By default, Python sets the current working directory to wherever your active file is located, so we don't have to worry about changing the directory as we did in R. You can still check and change it with the getcwd() and chdir() functions from os, respectively.

Python follows the same general rules about relative and absolute file paths as R. Let's see if we can locate *test-file.csv* in the repository using the isfile() function, which is in the path submodule of os:

```
In [24]: os.path.isfile('test-file.csv')
```

```
Out[24]: True
```

Now we'd like to locate that file as contained in the *test-folder* subfolder.

```
In [25]: os.path.isfile('test-folder/test-file.csv')
```

```
Out[25]: True
```

Next, try putting a copy of this file in the folder one up from your current location. You should be able to locate it with this code:

```
In [26]: os.path.isfile('../test-file.csv')
```

```
Out[26]: True
```

Like with R, you'll most commonly read data in from an external source to operate on it in Python, and this source can be nearly anything imaginable. pandas includes functions to read data from, among others, both *.xlsx* and *.csv* files into DataFrames. To demonstrate, we'll read in our reliable *star.xlsx* and *districts.csv* datasets from the book repository. The read_excel() function is used to read Excel workbooks:

```
In [27]: star = pd.read_excel('datasets/star/star.xlsx')
         star.head()
```

```
Out[27]:
   tmathssk  treadssk             classk  totexpk   sex freelunk   race
0       473       447        small.class        7  girl       no  white
1       536       450        small.class       21  girl       no  black
2       463       439  regular.with.aide        0   boy      yes  black
3       559       448            regular       16   boy       no  white
4       489       447        small.class        5   boy      yes  white

   schidkn
0       63
1       20
2       19
3       69
4       79
```

Similarly, we can use pandas to read in *.csv* files with the read_csv() function:

```
In [28]: districts = pd.read_csv('datasets/star/districts.csv')
         districts.head()
```

```
Out[28]:
           schidkn       school_name         county
      0          1            Rosalia    New Liberty
      1          2    Montgomeryville         Topton
      2          3               Davy       Wahpeton
      3          4           Steelton      Palestine
      4          6         Tolchester        Sattley
```

If you'd like to read in other Excel file types or specific ranges and worksheets, for example, check the pandas documentation.

Exploring a DataFrame

Let's continue to size up the *star* DataFrame. The info() method will tell us some important things, such as its dimensions and types of columns:

```
In [29]: star.info()

         <class 'pandas.core.frame.DataFrame'>
         RangeIndex: 5748 entries, 0 to 5747
         Data columns (total 8 columns):
          #   Column    Non-Null Count   Dtype
         ---  ------    --------------   -----
          0   tmathssk  5748 non-null    int64
          1   treadssk  5748 non-null    int64
          2   classk    5748 non-null    object
          3   totexpk   5748 non-null    int64
          4   sex       5748 non-null    object
          5   freelunk  5748 non-null    object
          6   race      5748 non-null    object
          7   schidkn   5748 non-null    int64
         dtypes: int64(4), object(4)
         memory usage: 359.4+ KB
```

We can retrieve descriptive statistics with the describe() method:

```
In [30]: star.describe()
```

```
Out[30]:
                  tmathssk      treadssk       totexpk        schidkn
      count    5748.000000   5748.000000   5748.000000    5748.000000
      mean      485.648051    436.742345      9.307411      39.836639
      std        47.771531     31.772857      5.767700      22.957552
      min       320.000000    315.000000      0.000000       1.000000
      25%       454.000000    414.000000      5.000000      20.000000
      50%       484.000000    433.000000      9.000000      39.000000
```

	tmathssk	treadssk		classk	totexpk	sex	\
75%	513.000000	453.000000		13.000000	60.000000		
max	626.000000	627.000000		27.000000	80.000000		

By default, `pandas` only includes descriptive statistics of numeric variables. We can override this with `include = 'all'`.

```
In [31]: star.describe(include = 'all')
```

```
Out[31]:
```

	tmathssk	treadssk		classk	totexpk	sex	\
count	5748.000000	5748.000000		5748	5748.000000	5748	
unique	NaN	NaN		3	NaN	2	
top	NaN	NaN	regular.with.aide		NaN	boy	
freq	NaN	NaN		2015	NaN	2954	
mean	485.648051	436.742345		NaN	9.307411	NaN	
std	47.771531	31.772857		NaN	5.767700	NaN	
min	320.000000	315.000000		NaN	0.000000	NaN	
25%	454.000000	414.000000		NaN	5.000000	NaN	
50%	484.000000	433.000000		NaN	9.000000	NaN	
75%	513.000000	453.000000		NaN	13.000000	NaN	
max	626.000000	627.000000		NaN	27.000000	NaN	

	freelunk	race	schidkn
count	5748	5748	5748.000000
unique	2	3	NaN
top	no	white	NaN
freq	2973	3869	NaN
mean	NaN	NaN	39.836639
std	NaN	NaN	22.957552
min	NaN	NaN	1.000000
25%	NaN	NaN	20.000000
50%	NaN	NaN	39.000000
75%	NaN	NaN	60.000000
max	NaN	NaN	80.000000

NaN is a special `pandas` value to indicate missing or unavailable data, such as the standard deviation of a categorical variable.

Indexing and Subsetting DataFrames

Let's return to the small *roster* DataFrame, accessing various elements by their row and column position. To index a DataFrame we can use the `iloc`, or *integer location*, method. The square bracket notation will look familiar to you, but this time we need to index by both row *and* column (again, both starting at zero). Let's demonstrate on the *roster* DataFrame we created earlier.

```
In [32]: # First row, first column of DataFrame
         roster.iloc[0, 0]
```

```
Out[32]: 'Jack'
```

It's possible to employ slicing here as well to capture multiple rows and columns:

```
In [33]: # Second through fourth rows, first through third columns
         roster.iloc[1:4, 0:3]

Out[33]:
     name height injury
  1   Jill     65   True
  2  Billy     68  False
  3  Susie     69  False
```

To index an entire column by name, we can use the related loc method. We'll leave a blank slice in the first index position to capture all rows, then name the column of interest:

```
In [34]: # Select all rows in the name column
         roster.loc[:, 'name']

Out[34]:
  0       Jack
  1       Jill
  2      Billy
  3      Susie
  4     Johnny
Name: name, dtype: object
```

Writing DataFrames

pandas also includes functions to write DataFrames to both *.csv* files and *.xlsx* workbooks with the write_csv() and write_xlsx() methods, respectively:

```
In [35]: roster.to_csv('output/roster-output-python.csv')
         roster.to_excel('output/roster-output-python.xlsx')
```

Conclusion

In a short time, you were able to progress all the way from single-element objects, to lists, to numpy arrays, then finally to pandas DataFrames. I hope you were able to see the evolution and linkage between these data structures while appreciating the added benefits of the packages introduced. The following chapters on Python will rely heavily on pandas, but you've seen here that pandas itself relies on numpy and the basic rules of Python, such as zero-based indexing.

Exercises

In this chapter, you learned how to work with a few different data structures and collection types in Python. The following exercises provide additional practice and insight on these topics:

1. Slice the following array so that you are left with the third through fifth elements.

   ```
   practice_array = ['I', 'am', 'having', 'fun', 'with', 'Python']
   ```

2. Load the `tips` DataFrame from `seaborn`.

   - Print some information about this DataFrame, such as the number of observations and each column's type.

   - Print the descriptive statistics for this DataFrame.

3. The book repository (*https://oreil.ly/RKmg0*) includes an *ais.xlsx* file in the *ais* subfolder of the *datasets* folder. Read it into Python as a DataFrame.

   - Print the first few rows of this DataFrame.

   - Write just the *sport* column of this DataFrame back to Excel as *sport.xlsx*.

## CHAPTER 12

Data Manipulation and Visualization in Python

In Chapter 8 you learned how to manipulate and visualize data, with heavy help from the tidyverse suite of packages. Here we'll demonstrate similar techniques on the same *star* dataset, this time in Python. In particular, we'll use pandas and seaborn to manipulate and visualize data, respectively. This isn't a comprehensive guide to what these modules, or Python, can do with data analysis. Instead, it's enough to get you exploring on your own.

As much as possible, I'll mirror the steps and perform the same operations that we did in Chapter 8. Because of this familiarity, I'll focus less on the whys of manipulating and visualizing data than I will on hows of doing it in Python. Let's load the necessary modules and get started with *star*. The third module, matplotlib, is new for you and will be used to complement our work in seaborn. It comes installed with Anaconda. Specifically, we'll be using the pyplot submodule, aliasing it as plt.

```
In [1]:  import pandas as pd
         import seaborn as sns
         import matplotlib.pyplot as plt

         star = pd.read_excel('datasets/star/star.xlsx')
         star.head()
Out[1]:
   tmathssk  treadssk            classk  totexpk  sex freelunk   race  \
0       473       447       small.class        7  girl       no  white
1       536       450       small.class       21  girl       no  black
2       463       439  regular.with.aide        0   boy      yes  black
3       559       448           regular       16   boy       no  white
4       489       447       small.class        5   boy      yes  white

   schidkn
```

```
0    63
1    20
2    19
3    69
4    79
```

Column-Wise Operations

In Chapter 11 you learned that `pandas` will attempt to convert one-dimensional data structures into Series. This seemingly trivial point will be quite important when selecting columns. Let's take a look at an example: say we *just* wanted to keep the *tmathssk* column from our DataFrame. We could do so using the familiar single-bracket notation, but this technically results in a Series, not a DataFrame:

```
In [2]:  math_scores = star['tmathssk']
         type(math_scores)

Out[2]: pandas.core.series.Series
```

It's probably better to keep this as a DataFrame if we aren't positive that we want *math_scores* to stay as a one-dimensional structure. To do so, we can use two sets of brackets instead of one:

```
In [3]: math_scores = star[['tmathssk']]
        type(math_scores)

Out[3]: pandas.core.frame.DataFrame
```

Following this pattern, we can keep only the desired columns in *star*. I'll use the `col` `umns` attribute to confirm.

```
In [4]:  star = star[['tmathssk','treadssk','classk','totexpk','schidkn']]
         star.columns

Out[4]: Index(['tmathssk', 'treadssk', 'classk',
               'totexpk', 'schidkn'], dtype='object')
```

Object-Oriented Programming in Python

So far you've seen methods and functions in Python. These are things that objects can *do*. Attributes, on the other hand, represent some *state* of an object itself. These are affixed to an object's name with a period; unlike methods, no parentheses are used. Attributes, functions, and methods are all elements of *object-oriented programming* (OOP), a paradigm meant to structure work into simple and reusable pieces of code. To learn more about how OOP works in Python, check out *Python in a Nutshell*, 3rd edition by Alex Martelli et al. (O'Reilly).

To drop specific columns, use the drop() method. drop() can be used to drop columns *or* rows, so we'll need to specify which by using the axis argument. In pandas, rows are axis 0 and columns axis 1, as Figure 12-1 demonstrates.

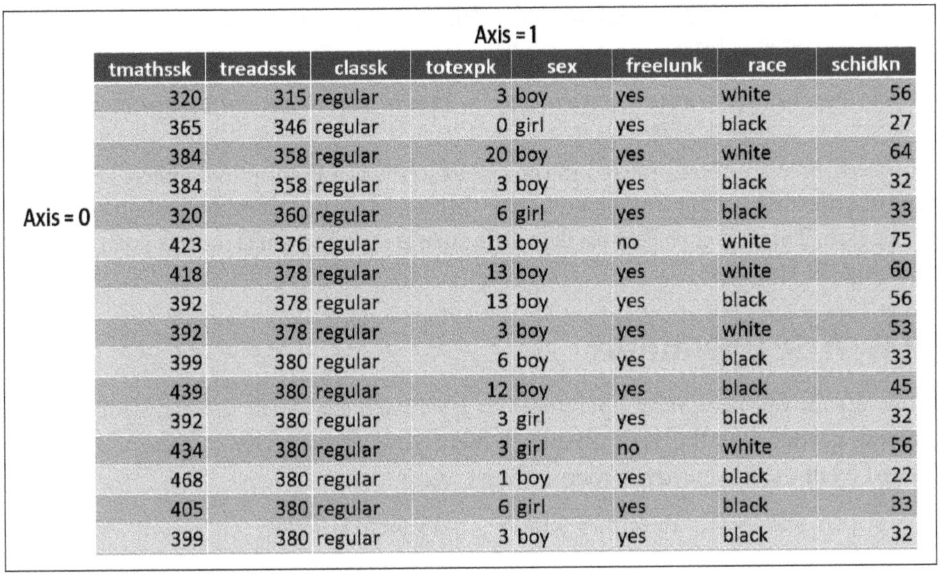

	tmathssk	treadssk	classk	totexpk	sex	freelunk	race	schidkn
				Axis = 1				
	320	315	regular	3	boy	yes	white	56
	365	346	regular	0	girl	yes	black	27
	384	358	regular	20	boy	yes	white	64
	384	358	regular	3	boy	yes	black	32
Axis = 0	320	360	regular	6	girl	yes	black	33
	423	376	regular	13	boy	no	white	75
	418	378	regular	13	boy	yes	white	60
	392	378	regular	13	boy	yes	black	56
	392	378	regular	3	boy	yes	white	53
	399	380	regular	6	boy	yes	black	33
	439	380	regular	12	boy	yes	black	45
	392	380	regular	3	girl	yes	black	32
	434	380	regular	3	girl	no	white	56
	468	380	regular	1	boy	yes	black	22
	405	380	regular	6	girl	yes	black	33
	399	380	regular	3	boy	yes	black	32

Figure 12-1. Axes of a pandas DataFrame

Here's how to drop the *schidkn* column:

```
In [5]: star = star.drop('schidkn', axis=1)
        star.columns

Out[5]: Index(['tmathssk', 'treadssk',
            'classk', 'totexpk'], dtype='object')
```

Let's now look at deriving new columns of a DataFrame. We can do that using bracket notation—this time, I *do* want the result to be a Series, as each column of a Data-Frame is actually a Series (just as each column of an R data frame is actually a vector). Here I'll calculate combined math and reading scores:

```
In [6]: star['new_column'] = star['tmathssk'] + star['treadssk']
        star.head()

Out[6]:
    tmathssk  treadssk             classk  totexpk  new_column
0       473       447         small.class        7         920
1       536       450         small.class       21         986
2       463       439   regular.with.aide        0         902
3       559       448             regular       16        1007
4       489       447         small.class        5         936
```

Again, *new_column* isn't a terribly descriptive variable name. Let's fix that with the rename() function. We'll use the columns argument and pass data to it in a format you're likely unfamiliar with:

```
In [7]: star = star.rename(columns = {'new_column':'ttl_score'})
        star.columns

Out[7]: Index(['tmathssk', 'treadssk', 'classk', 'totexpk', 'ttl_score'],
            dtype='object')
```

The curly bracket notation used in the last example is a Python *dictionary*. Dictionaries are collections of *key-value* pairs, with the key and value of each element separated by a colon. This is a core Python data structure and one to check out as you continue learning the language.

Row-Wise Operations

Now let's move to common operations by row. We'll start with sorting, which can be done in pandas with the sort_values() method. We'll pass a list of columns we want to sort by in their respective order to the by argument:

```
In [8]: star.sort_values(by=['classk', 'tmathssk']).head()

Out[8]:
      tmathssk  treadssk  classk  totexpk  ttl_score
309        320       360  regular       6        680
1470       320       315  regular       3        635
2326       339       388  regular       6        727
2820       354       398  regular       6        752
4925       354       391  regular       8        745
```

By default, all columns are sorted ascendingly. To modify that, we can include another argument, ascending, which will contain a list of True/False flags. Let's sort *star* by class size (*classk*) ascending and math score (*treadssk*) descending. Because we're not assigning this output back to *star*, this sorting is not permanent to the dataset.

```
In [9]: # Sort by class size ascending and math score descending
        star.sort_values(by=['classk', 'tmathssk'],
            ascending=[True, False]).head()

Out[9]:
      tmathssk  treadssk  classk  totexpk  ttl_score
724        626       474  regular      15       1100
1466       626       554  regular      11       1180
1634       626       580  regular      15       1206
2476       626       538  regular      20       1164
2495       626       522  regular       7       1148
```

To filter a DataFrame, we'll first use conditional logic to create a Series of `True`/`False` flags indicating whether each row meets some criteria. We'll then keep only the rows in the DataFrame where records in the Series are flagged as `True`. For example, let's keep only the records where `classk` is equal to `small.class`.

```
In [10]: small_class = star['classk'] == 'small.class'
         small_class.head()

Out[10]:
         0     True
         1     True
         2     False
         3     False
         4     True
         Name: classk, dtype: bool
```

We can now filter by this resulting Series by using brackets. We can confirm the number of rows and columns in our new DataFrame with the `shape` attribute:

```
In [11]: star_filtered = star[small_class]
         star_filtered.shape

Out[11]: (1733, 5)
```

`star_filtered` will contain fewer rows than *star*, but the same number of columns:

```
In [12]: star.shape

Out[12]: (5748, 5)
```

Let's try one more: we'll find the records where `treadssk` is at least 500:

```
In [13]: star_filtered = star[star['treadssk'] >= 500]
         star_filtered.shape

Out[13]: (233, 5)
```

It's also possible to filter by multiple conditions using and/or statements. Just like in R, & and | indicate "and" and "or" in Python, respectively. Let's pass both of the previous criteria into one statement by placing each in parentheses, connecting them with &:

```
In [14]: # Find all records with reading score at least 500 and in small class
         star_filtered = star[(star['treadssk'] >= 500) &
                     (star['classk'] == 'small.class')]
         star_filtered.shape

Out[14]: (84, 5)
```

Aggregating and Joining Data

To group observations in a DataFrame, we'll use the `groupby()` method. If we print `star_grouped`, you'll see it's a `DataFrameGroupBy` object:

```
In [15]: star_grouped = star.groupby('classk')
         star_grouped

Out[15]: <pandas.core.groupby.generic.DataFrameGroupBy
             object at 0x000001EFD8DFF388>
```

We can now choose other fields to aggregate this grouped DataFrame by. Table 12-1 lists some common aggregation methods.

Table 12-1. Helpful aggregation functions in pandas

Method	Aggregation type
sum()	Sum
count()	Count values
mean()	Average
max()	Highest value
min()	Lowest value
std()	Standard deviation

The following gives us the average math score for each class size:

```
In [16]: star_grouped[['tmathssk']].mean()

Out[16]:
                        tmathssk
         classk
         regular             483.261000
         regular.with.aide   483.009926
         small.class         491.470283
```

Now we'll find the highest total score for each year of teacher experience. Because this would return quite a few rows, I will include the `head()` method to get just a few. This practice of adding multiple methods to the same command is called method *chaining*:

```
In [17]: star.groupby('totexpk')[['ttl_score']].max().head()

Out[17]:
                ttl_score
         totexpk
         0           1171
         1           1133
         2           1091
         3           1203
         4           1229
```

Chapter 8 reviewed the similarities and differences between Excel's VLOOKUP() and a left outer join. I'll read in a fresh copy of *star* as well as *districts*; let's use pandas to join these datasets. We'll use the merge() method to "look up" data from *school-districts* into *star*. By setting the how argument to left, we'll specify a left outer join, which again is the join type most similar to VLOOKUP():

```
In [18]: star = pd.read_excel('datasets/star/star.xlsx')
         districts = pd.read_csv('datasets/star/districts.csv')
         star.merge(districts, how='left').head()
```

```
Out[18]:
     tmathssk  treadssk             classk  totexpk  sex freelunk   race  \
0         473       447        small.class        7  girl       no  white
1         536       450        small.class       21  girl       no  black
2         463       439  regular.with.aide        0   boy      yes  black
3         559       448            regular       16   boy       no  white
4         489       447        small.class        5   boy      yes  white

     schidkn     school_name          county
0         63       Ridgeville     New Liberty
1         20    South Heights         Selmont
2         19        Bunnlevel         Sattley
3         69            Hokah      Gallipolis
4         79    Lake Mathews   Sugar Mountain
```

Python, like R, is quite intuitive about joining data: it knew by default to merge on *schidkn* and brought in both *school_name* and *county*.

Reshaping Data

Let's take a look at widening and lengthening a dataset in Python, again using pandas. To start, we can use the melt() function to combine *tmathssk* and *treadssk* into one column. To do this, I'll specify the DataFrame to manipulate with the frame argument, which variable to use as a unique identifier with id_vars, and which variables to melt into a single column with value_vars. I'll also specify what to name the resulting value and label variables with value_name and var_name, respectively:

```
In [19]: star_pivot = pd.melt(frame=star, id_vars = 'schidkn',
             value_vars=['tmathssk', 'treadssk'], value_name='score',
             var_name='test_type')
         star_pivot.head()
```

```
Out[19]:
     schidkn test_type  score
0         63  tmathssk    473
1         20  tmathssk    536
2         19  tmathssk    463
3         69  tmathssk    559
4         79  tmathssk    489
```

How about renaming *tmathssk* and *treadssk* as *math* and *reading*, respectively? To do this, I'll use a Python dictionary to set up an object called mapping, which serves as something like a "lookup table" to recode the values. I'll pass this into the map() method which will recode *test_type*. I'll also use the unique() method to confirm that only *math* and *reading* are now found in *test_type*:

```
In [20]: # Rename records in `test_type`
         mapping = {'tmathssk':'math','treadssk':'reading'}
         star_pivot['test_type'] = star_pivot['test_type'].map(mapping)

         # Find unique values in test_type
         star_pivot['test_type'].unique()

Out[20]: array(['math', 'reading'], dtype=object)
```

To widen *star_pivot* back into separate *math* and *reading* columns, I'll use the pivot_table() method. First I'll specify what variable to index by with the index argument, then which variables contain the labels and values to pivot from with the columns and values arguments, respectively.

It's possible in pandas to set unique index columns; by default, pivot_table() will set whatever variables you've included in the index argument there. To override this, I'll use the reset_index() method. To learn more about custom indexing in pandas, along with countless other data manipulation and analysis techniques we couldn't cover here, check out *Python for Data Analysis*, 2nd edition by Wes McKinney (O'Reilly).

```
In [21]: star_pivot.pivot_table(index='schidkn',
             columns='test_type', values='score').reset_index()

Out[21]:
         test_type  schidkn        math      reading
         0                1  492.272727   443.848485
         1                2  450.576923   407.153846
         2                3  491.452632   441.000000
         3                4  467.689655   421.620690
         4                5  460.084746   427.593220
         ..             ...         ...          ...
         74              75  504.329268   440.036585
         75              76  490.260417   431.666667
         76              78  468.457627   417.983051
         77              79  490.500000   434.451613
         78              80  490.037037   442.537037

         [79 rows x 3 columns]
```

Data Visualization

Let's now briefly touch on data visualization in Python, specifically using the seaborn package. seaborn works especially well for statistical analysis and with pandas Data-Frames, so it's a great choice here. Similarly to how pandas is built on top of numpy, seaborn leverages features of another popular Python plotting package, matplotlib.

seaborn includes many functions to build different plot types. We'll modify the arguments within these functions to specify which dataset to plot, which variables go along the x- and y-axes, which colors to use, and so on. Let's get started by visualizing the count of observations for each level of *classk*, which we can do with the countplot() function.

Our dataset is *star*, which we'll specify with the data argument. To place our levels of *classk* along the x-axis we'll use the x argument. This results in the countplot shown in Figure 12-2:

```
In [22]: sns.countplot(x='classk', data=star)
```

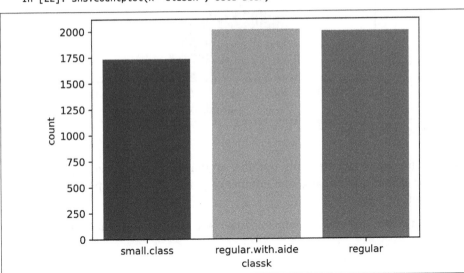

Figure 12-2. Countplot

Now for a histogram of *treadssk*, we'll use the displot() function. Again, we'll specify x and data. Figure 12-3 shows the result:

```
In [23]: sns.displot(x='treadssk', data=star)
```

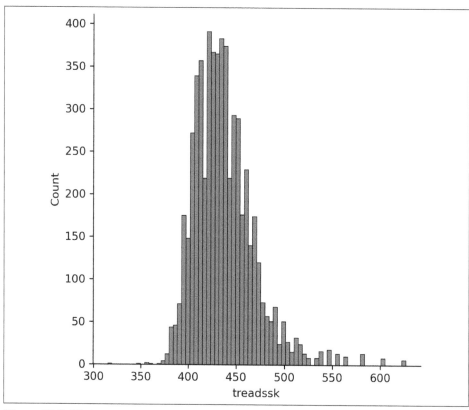

Figure 12-3. Histogram

seaborn functions include many optional arguments to customize a plot's appearance. For example, let's change the number of bins to 25 and the plot color to pink. This results in Figure 12-4:

```
In [24]: sns.displot(x='treadssk', data=star, bins=25, color='pink')
```

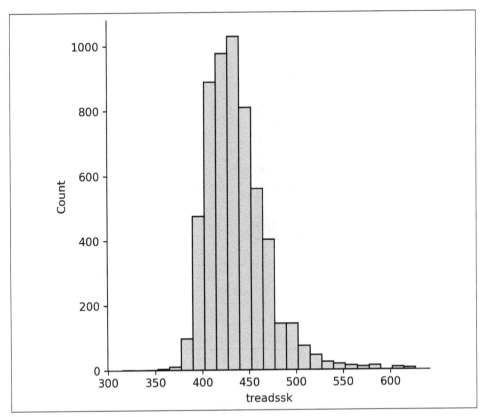

Figure 12-4. Custom histogram

To make a boxplot, use the boxplot() function as in Figure 12-5:

```
In [25]: sns.boxplot(x='treadssk', data=star)
```

In any of these cases so far, we could've "flipped" the plot by instead including the variable of interest in the y argument. Let's try it with our boxplot, and we'll get what's shown in Figure 12-6 as output:

```
In [26]: sns.boxplot(y='treadssk', data=star)
```

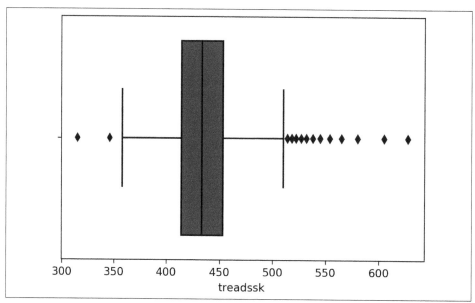

Figure 12-5. Boxplot

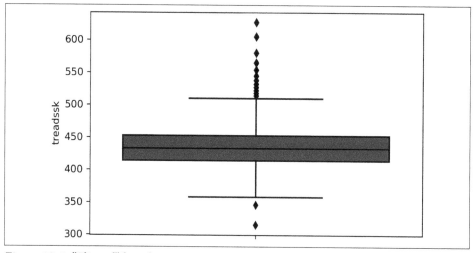

Figure 12-6. "Flipped" boxplot

To make a boxplot for each level of class size, we'll include an additional argument to plot *classk* along the x-axis, giving us the boxplot by group depicted in Figure 12-7:

```
In [27]: sns.boxplot(x='classk', y='treadssk', data=star)
```

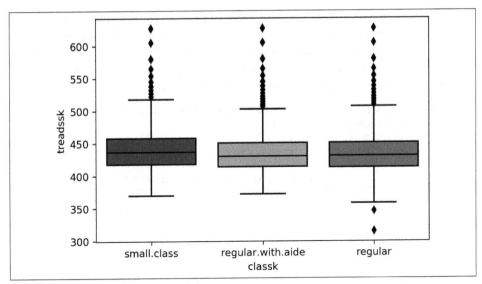

Figure 12-7. Boxplot by group

Now let's use the `scatterplot()` function to plot the relationship of *tmathssk* on the x-axis and *treadssk* on the y. Figure 12-8 is the result:

```
In [28]: sns.scatterplot(x='tmathssk', y='treadssk', data=star)
```

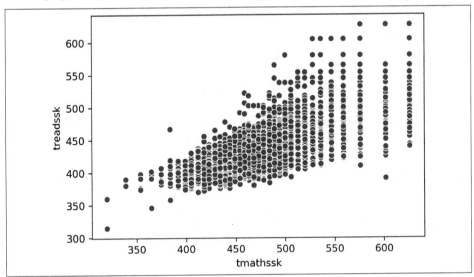

Figure 12-8. Scatterplot

Let's say we wanted to share this plot with an outside audience, who may not be familiar with what *treadssk* and *tmathssk* are. We can add more helpful labels to this chart by borrowing features from `matplotlib.pyplot`. We'll run the same `scatter plot()` function as before, but this time we'll also call in functions from `pyplot` to add custom x- and y-axis labels, as well as a chart title. This results in Figure 12-9:

```
In [29]: sns.scatterplot(x='tmathssk', y='treadssk', data=star)
         plt.xlabel('Math score')
         plt.ylabel('Reading score')
         plt.title('Math score versus reading score')
```

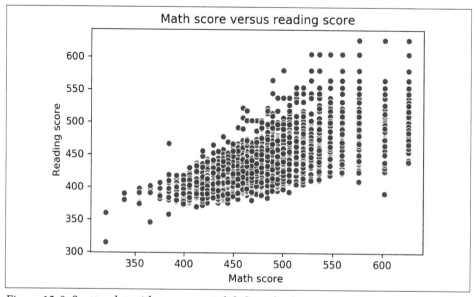

Figure 12-9. Scatterplot with custom axis labels and title

seaborn includes many more features for building visually appealing data visualizations. To learn more, check out the official documentation (*https://oreil.ly/2joMU*).

Conclusion

There's so much more that `pandas` and `seaborn` can do, but this is enough to get you started with the true task at hand: to explore and test relationships in data. That will be the focus of Chapter 13.

Exercises

The book repository (*https://oreil.ly/hFEOG*) has two files in the *census* subfolder of datasets, *census.csv* and *census-divisions.csv*. Read these into Python and do the following:

1. Sort the data by region ascending, division ascending and population descending. (You will need to combine datasets to do this.) Write the results to an Excel worksheet.

2. Drop the postal code field from your merged dataset.

3. Create a new column, *density*, that is a calculation of population divided by land area.

4. Visualize the relationship between land area and population for all observations in 2015.

5. Find the total population for each region in 2015.

6. Create a table containing state names and populations, with the population for each year 2010–2015 kept in an individual column.

## CHAPTER 13

Capstone: Python for Data Analytics

At the end of Chapter 8 you extended what you learned about R to explore and test relationships in the *mpg* dataset. We'll do the same in this chapter, using Python. We've conducted the same work in Excel and R, so I'll focus less on the whys of our analysis in favor of the hows of doing it in Python.

To get started, let's call in all the necessary modules. Some of these are new: from scipy, we'll import the `stats` submodule. To do this, we'll use the `from` keyword to tell Python what module to look for, then the usual `import` keyword to choose a submodule. As the name suggests, we'll use the `stats` submodule of `scipy` to conduct our statistical analysis. We'll also be using a new package called `sklearn`, or *scikit-learn*, to validate our model on a train/test split. This package has become a dominant resource for machine learning and also comes installed with Anaconda.

```
In [1]: import pandas as pd
        import seaborn as sns
        import matplotlib.pyplot as plt
        from scipy import stats
        from sklearn import linear_model
        from sklearn import model_selection
        from sklearn import metrics
```

With the `usecols` argument of `read_csv()` we can specify which columns to read into the DataFrame:

```
In [2]: mpg = pd.read_csv('datasets/mpg/mpg.csv',usecols=
        ['mpg','weight','horsepower','origin','cylinders'])
        mpg.head()

Out[2]:
     mpg  cylinders  horsepower  weight origin
0   18.0          8         130    3504    USA
1   15.0          8         165    3693    USA
```

```
2  18.0        8        150    3436    USA
3  16.0        8        150    3433    USA
4  17.0        8        140    3449    USA
```

Exploratory Data Analysis

Let's start with the descriptive statistics:

```
In[3]: mpg.describe()

Out[3]:
              mpg    cylinders  horsepower      weight
count  392.000000  392.000000  392.000000  392.000000
mean    23.445918    5.471939  104.469388  2977.584184
std      7.805007    1.705783   38.491160   849.402560
min      9.000000    3.000000   46.000000  1613.000000
25%     17.000000    4.000000   75.000000  2225.250000
50%     22.750000    4.000000   93.500000  2803.500000
75%     29.000000    8.000000  126.000000  3614.750000
max     46.600000    8.000000  230.000000  5140.000000
```

Because *origin* is a categorical variable, by default it doesn't show up as part of describe(). Let's explore this variable instead with a frequency table. This can be done in pandas with the crosstab() function. First, we'll specify what data to place on the index: *origin*. We'll get a count for each level by setting the columns argument to count:

```
In [4]: pd.crosstab(index=mpg['origin'], columns='count')

Out[4]:
col_0   count
origin
Asia       79
Europe     68
USA       245
```

To make a two-way frequency table, we can instead set columns to another categorical variable, such as cylinders:

```
In [5]: pd.crosstab(index=mpg['origin'], columns=mpg['cylinders'])

Out[5]:
cylinders  3   4  5   6    8
origin
Asia       4  69  0   6    0
Europe     0  61  3   4    0
USA        0  69  0  73  103
```

Next, let's retrieve descriptive statistics for *mpg* by each level of *origin*. I'll do this by chaining together two methods, then subsetting the results:

```
In[6]: mpg.groupby('origin').describe()['mpg']
```

```
Out[6]:
        count       mean        std   min    25%   50%     75%   max
origin
Asia     79.0  30.450633   6.090048  18.0  25.70  31.6  34.050  46.6
Europe   68.0  27.602941   6.580182  16.2  23.75  26.0  30.125  44.3
USA     245.0  20.033469   6.440384   9.0  15.00  18.5  24.000  39.0
```

We can also visualize the overall distribution of *mpg*, as in Figure 13-1:

```
In[7]: sns.displot(data=mpg, x='mpg')
```

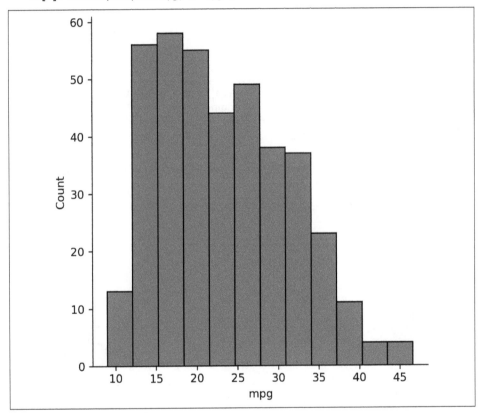

Figure 13-1. Histogram of mpg

Now let's make a boxplot as in Figure 13-2 comparing the distribution of *mpg* across each level of *origin*:

```
In[8]: sns.boxplot(x='origin', y='mpg', data=mpg, color='pink')
```

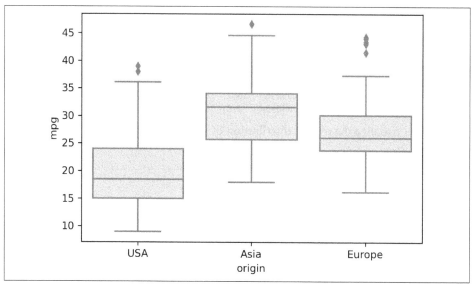

Figure 13-2. Boxplot of mpg by origin

Alternatively, we can set the `col` argument of `displot()` to `origin` to create faceted histograms, such as in Figure 13-3:

```
In[9]: sns.displot(data=mpg, x="mpg", col="origin")
```

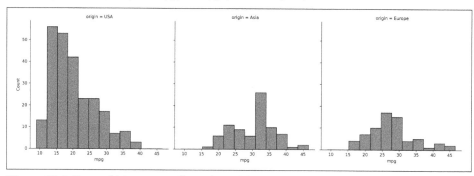

Figure 13-3. Faceted histogram of mpg by origin

Hypothesis Testing

Let's again test for a difference in mileage between American and European cars. For ease of analysis, we'll split the observations in each group into their own DataFrames.

```
In[10]: usa_cars = mpg[mpg['origin']=='USA']
        europe_cars = mpg[mpg['origin']=='Europe']
```

Independent Samples T-test

We can now use the `ttest_ind()` function from `scipy.stats` to conduct the t-test. This function expects two `numpy` arrays as arguments; `pandas` Series also work:

```
In[11]: stats.ttest_ind(usa_cars['mpg'], europe_cars['mpg'])

Out[11]: Ttest_indResult(statistic=-8.534455914399228,
         pvalue=6.306531719750568e-16)
```

Unfortunately, the output here is rather scarce: while it does include the p-value, it doesn't include the confidence interval. To run a t-test with more output, check out the `researchpy` module.

Let's move on to analyzing our continuous variables. We'll start with a correlation matrix. We can use the `corr()` method from `pandas`, including only the relevant variables:

```
In[12]: mpg[['mpg','horsepower','weight']].corr()

Out[12]:
                 mpg  horsepower    weight
mpg         1.000000   -0.778427 -0.832244
horsepower -0.778427    1.000000  0.864538
weight     -0.832244    0.864538  1.000000
```

Next, let's visualize the relationship between *weight* and *mpg* with a scatterplot as shown in Figure 13-4:

```
In[13]: sns.scatterplot(x='weight', y='mpg', data=mpg)
        plt.title('Relationship between weight and mileage')
```

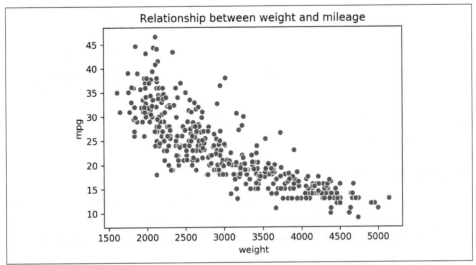

Figure 13-4. Scatterplot of mpg by weight

Alternatively, we could produce scatterplots across all pairs of our dataset with the `pairplot()` function from `seaborn`. Histograms of each variable are included along the diagonal, as seen in Figure 13-5:

```
In[14]: sns.pairplot(mpg[['mpg','horsepower','weight']])
```

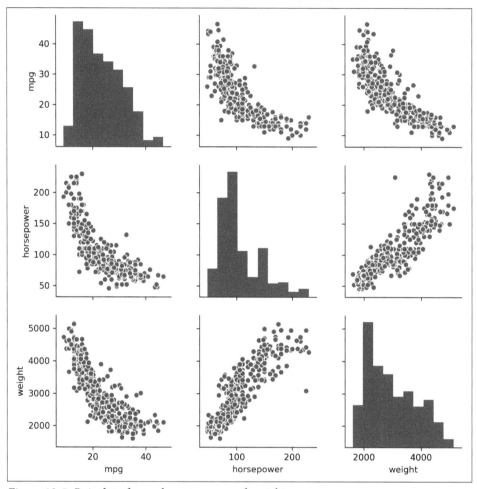

Figure 13-5. Pairplot of mpg, horsepower, and weight

Linear Regression

Now it's time for a linear regression. To do this, we'll use `linregress()` from `scipy`, which also looks for two `numpy` arrays or `pandas` Series. We'll specify which variable is our independent and dependent variable with the x and y arguments, respectively:

```
In[15]: # Linear regression of weight on mpg
        stats.linregress(x=mpg['weight'], y=mpg['mpg'])
```

```
Out[15]: LinregressResult(slope=-0.007647342535779578,
    intercept=46.21652454901758, rvalue=-0.8322442148315754,
    pvalue=6.015296051435726e-102, stderr=0.0002579632782734318)
```

Again, you'll see that some of the output you may be used to is missing here. *Be careful*: the rvalue included is the *correlation coefficient*, not R-square. For a richer linear regression output, check out the statsmodels module.

Last but not least, let's overlay our regression line to a scatterplot. seaborn has a separate function to do just that: regplot(). As usual, we'll specify our independent and dependent variables, and where to get the data. This results in Figure 13-6:

```
In[16]: # Fit regression line to scatterplot
    sns.regplot(x="weight", y="mpg", data=mpg)
    plt.xlabel('Weight (lbs)')
    plt.ylabel('Mileage (mpg)')
    plt.title('Relationship between weight and mileage')
```

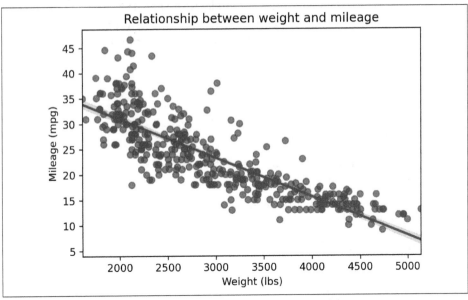

Figure 13-6. Scatterplot with fit regression line of mpg by weight

Train/Test Split and Validation

At the end of Chapter 9 you learned how to apply a train/test split when building a linear regression model in R.

We will use the train_test_split() function to split our dataset into *four* DataFrames: not just by training and testing but also independent and dependent variables. We'll pass in a DataFrame containing our independent variable first, then one containing the dependent variable. Using the `random_state` argument, we'll seed the random number generator so the results remain consistent for this example:

```
In[17]: X_train, X_test, y_train, y_test =
        model_selection.train_test_split(mpg[['weight']], mpg[['mpg']],
        random_state=1234)
```

By default, the data is split 75/25 between training and testing subsets:

```
In[18]: y_train.shape
```

```
Out[18]: (294, 1)
```

```
In[19]: y_test.shape
```

```
Out[19]: (98, 1)
```

Now, let's fit the model to the training data. First we'll specify the linear model with `LinearRegression()`, then we'll train the model with `regr.fit()`. To get the predicted values for the test dataset, we can use `predict()`. This results in a numpy array, not a pandas DataFrame, so the `head()` method won't work to print the first few rows. We can, however, slice it:

```
In[20]: # Create linear regression object
        regr = linear_model.LinearRegression()

        # Train the model using the training sets
        regr.fit(X_train, y_train)

        # Make predictions using the testing set
        y_pred = regr.predict(X_test)

        # Print first five observations
        y_pred[:5]
```

```
Out[20]: array([[14.86634263],
        [23.48793632],
        [26.2781699 ],
        [27.69989655],
        [29.05319785]])
```

The `coef_` attribute returns the coefficient of our test model:

```
In[21]: regr.coef_
```

```
Out[21]: array([[-0.00760282]])
```

To get more information about the model, such as the coefficient p-values or R-squared, try fitting it with the `statsmodels` package.

For now, we'll evaluate the performance of the model on our test data, this time using the `metrics` submodule of `sklearn`. We'll pass in our actual and predicted values to the `r2_score()` and `mean_squared_error()` functions, which will return the R-squared and RMSE, respectively.

```
In[22]:  metrics.r2_score(y_test, y_pred)

Out[22]: 0.6811923996681357


In[23]:  metrics.mean_squared_error(y_test, y_pred)

Out[23]: 21.63348076436662
```

Conclusion

The usual caveat applies to this chapter: we've just scratched the surface of what analysis is possible on this or any other dataset. But I hope you feel you've hit your stride on working with data in Python.

Exercises

Take another look at the *ais* dataset, this time using Python. Read the Excel workbook in from the book repository (*https://oreil.ly/dsZDM*) and complete the following. You should be pretty comfortable with this analysis by now.

1. Visualize the distribution of red blood cell count (*rcc*) by sex (*sex*).
2. Is there a significant difference in red blood cell count between the two groups of sex?
3. Produce a correlation matrix of the relevant variables in this dataset.
4. Visualize the relationship of height (*ht*) and weight (*wt*).
5. Regress *ht* on *wt*. Find the equation of the fit regression line. Is there a significant relationship?
6. Split your regression model into training and testing subsets. What is the R-squared and RMSE on your test model?

## CHAPTER 14

Conclusion and Next Steps

In the Preface, I stated the following learning objective:

> By the end of this book, you should be able to *conduct exploratory data analysis and hypothesis testing using a programming language.*

I sincerely hope you feel this objective has been met, and that you are confident to advance into further areas of analytics. To end this leg of your analytics journey, I'd like to share some topics to help round out and expand upon what you now know.

Further Slices of the Stack

Chapter 5 covered four major categories of software applications used in data analytics: spreadsheets, programming languages, databases, and BI tools. Because of our focus on the statistically based elements of analytics, we emphasized the first two slices of the stack. Refer back to that chapter on ideas for how the other slices tie in, and what to learn about them.

Research Design and Business Experiments

You learned in Chapter 3 that sound data analysis can only follow from sound data *collection*: as the saying goes, "garbage in, garbage out." In this book, we've assumed our data was collected accurately, was the right data for our analysis, and contained a representative sample. And we've been working with well-known datasets often taken from peer-reviewed research, so this is a safe assumption.

But often you can't be so sure about your data; you may responsible for collecting *and* analyzing it. It's worth learning more, then, about research *design* and *methods*. This field can get quite sophisticated and academic, but it's found practical applications in the field of business experiments. Check out Stefan H. Thomke's *Experimentation*

Works: The Surprising Power of Business Experiments (Harvard Business Review Press) for an overview of how and why to apply sound research methods to business.

Further Statistical Methods

As Chapter 4 mentioned, we've only scratched the surface of the types of statistical tests available, although many of them rest on the same framework of hypothesis testing covered in Chapter 3.

For a conceptual overview other statistical methods, check out Sarah Boslaugh's *Statistics in a Nutshell*, 2nd edition (O'Reilly). Then, head to *Practical Statistics for Data Scientists*, 2nd Edition (O'Reilly) by Peter Bruce et al. to apply them using R and Python. As its title suggests, the latter book straddles the line between statistics and data science.

Data Science and Machine Learning

Chapter 5 reviewed the differences between statistics, data analytics, and data science and concluded that although there are differences in methods, more unites these fields than divides them.

If you are keenly interested in data science and machine learning, focus your learning efforts on R and Python, with some SQL and database knowledge too. To see how R is used in data science, check out *R for Data Science* by Hadley Wickham and Garrett Grolemund (O'Reilly). For Python, check out *Hands-On Machine Learning with Scikit-Learn, Keras, and TensorFlow*, 2nd edition by Aurélien Géron (O'Reilly).

Version Control

Chapter 5 also mentioned the importance of reproducibility. Let's look at a key application in that fight. You've likely run into a set of files like the following before:

- *proposal.txt*
- *proposal-v2.txt*
- *proposal-Feb23.txt*
- *proposal-final.txt*
- *proposal-FINAL-final.txt*

Maybe one user created *proposal-v2.txt*; another started *proposal-Feb23.txt*. Then there's the difference between *proposal-final.txt* and *proposal-FINAL-final.txt* to contend with. It can be quite difficult to sort out which is the "main" copy, and how to reconstruct and migrate all changes to that copy while keeping a record of who contributed what.

A *version control system* can come to the rescue here. This is a way to track projects over time, such as the contributions and changes made by different users. Version control is a game changer for collaboration and tracking revisions but has a relatively steep learning curve.

Git is a dominant version control system that is quite popular among data scientists, software engineers, and other technical professionals. In particular, they often use the cloud-based hosting service GitHub to manage Git projects. For an overview of Git and GitHub, check out *Version Control with Git*, 2nd edition by Jon Loeliger and Matthew McCullough (O'Reilly). For a look at how to pair Git and GitHub with R and RStudio, check out the online resource *Happy Git and GitHub for the useR* (*https:// happygitwithr.com*) by Jenny Bryan et al. At this time Git and other version control systems are relatively uncommon in data analytics workflows, but they are growing in popularity due in part to the growing demand for reproducibility.

Ethics

From recording and collecting it to analyzing and modeling it, data is surrounded by ethical concerns. In Chapter 3 you learned about statistical bias: especially in a machine learning context, it's possible that a model could begin to discriminate against groups of people in unjust or illegal ways. If data is being collected about individuals, consideration should be given to those individuals' privacy and consent.

Ethics hasn't always been a priority in data analytics and data science. Fortunately, the tide appears to be turning here, and will only continue with sustained community support. For a brief guide on how to incorporate ethical standards into working with data, check out *Ethics and Data Science* by Mike Loukides et al. (O'Reilly).

Go Forth and Data How You Please

I'm often asked which of these tools one should focus on given employer demand and trending popularity. My answer: take some time to find out what you like, and let those interests shape your learning path rather than trying to tailor it toward the "next big thing" in analytics tools. These skills are *all* valuable. More important than any one analytics tool is the ability to contextualize and pair those tools, which requires exposure to a broad set of applications. But you can't become an expert in everything. The best learning strategy will resemble a "T" shape: a wide exposure to various data tools, with relatively deeper knowledge in a handful of them.

Parting Words

Take a moment to look back at everything you've accomplished with this book: you should be proud. But don't linger: there's so much more to learn, and it won't take long in your learning journey to realize what a tip of the iceberg this book represented to you. Here's your end-of-chapter and end-of-book exercise: get out there, keep learning, and continue advancing into analytics.

Index

boxplot() function, Python, 197
boxplots
    for distribution by quartiles, 24-26
    with Python, 197-199, 205
    with R, 139-140, 148
    with Excel, 24-26
business analytics, 66, 80
business intelligence (BI) platforms, 82, 84, 86

C

c() function, R, 110, 112, 113, 121
Calculate Now, Excel, 31
calculators, applications as, xii, 98, 107, 166, 173
case-sensitive languages, 99, 168
categorical variables
    about, 6-8, 10, 64
    descriptive statistics for, 19-22
    in Excel, 12-14
    in multiple linear regression, 75
    in Python, 204
    in R, 114, 146, 158
causation, correlation and, 61, 76
cause and effect, 45-46
cell comments, 98
central limit theorem (CLT), 37-40, 46, 51
central tendency, measuring, 15-17, 26
chaining, method, Python, 192, 204
Chart Elements, Excel, 14, 65, 70
charts (see visualization of data)
chdir() function, Python, 181
classification models, 158
classifying variables, 5-11, 45
CLT (central limit theorem), 37-40, 46, 51
Clustered Columns, Excel, 14, 23, 29, 32
Code cell format, Jupyter, 165, 166
coefficient of determination (R-squared), 74, 157
coefficient of slope, 68, 71
coefficient, correlation (see correlation coefficient)
coef_attribute, Python, 210
coercing of data elements, 111, 177
col argument, Python, 206
collecting data, 42-43, 57, 213
collection object types, Python, 175
colon (:), R operator, 95, 112, 120, 126
columns
    creating new, 4, 126, 189

data manipulation operations for, 4, 124-126, 135, 188-190, 193-195
    in DataFrames, Python, 180, 184
    dropping, 124-125, 189
    filtering, 128
    keeping, 188
    as variables in data analytics, 5-11
columns argument, Python, 194, 204
columns attribute, Python, 188
combining (joining) data, 131-132, 192-193
command-line interfaces, Python, 163, 171
comments in code, 98, 167
comparison operators in R and Python, 99, 168
Comprehensive R Archive Network (CRAN), 93, 103-105, 107
conda install, Python, 172
conda update commands, Python, 172
conditional probability, 28
confidence intervals
    defined, 51
    in Excel, 51-56, 58, 69, 73
    formula (Excel) for, 52
    for linear regression, 69, 73
    in Python, 207
    in R, 151, 155
confirmatory data analysis (see hypothesis testing)
Console/Terminal, RStudio, 94
continuous probability distributions, 32-40
continuous variables
    about, 6, 8
    classifying, 10
    correlation analysis with, 62-66
    descriptive statistics for, 19, 22-24
    relationships between, 46, 62-65, 68, 75, 151, 207
cor() function, R, 151
corr() method, Python, 207
CORREL() function, Excel, 64
correlation and regression, 61-77
    causation, implication of, 61, 76
    correlation calculation and analysis, 62-66
    exercises, 77
    linear regression, 67-75
    mpg dataset example for, 63-66, 69-75
    Pearson, 62
    spurious relationships, 75
correlation coefficient
    with Excel, 64-66

null, rejection of, 50, 54, 61, 73, 151
numbers in R, 101
NumPy, Python, 176, 177
   (see also arrays, Python NumPy)

O

object-oriented programming (OOP), 188
objects
   naming, 101, 168
   in Python, 168-170, 175
   in R, 97, 100-102, 105
   structure of, 102, 110
   types of, 101, 169, 175
   versus variables, 100
observations in data analytics
   in calculations, 15, 17
   distributions of, 21, 24
   outliers, 25
   quantitative variables as, 8
   as rows in dataset, 5
OLS (ordinary least squares), 73
one-based indexing, 178
one-dimensional data structures, 176, 179, 188
one-way frequency tables, 12, 146
OOP (object-oriented programming), 188
open source software, 87, 94, 161, 163
operators
   in Python, 166, 168
   in R, 98, 99
order of operations, 98, 167
ordering, intrinsic, 8
ordinal variables, 6, 8
ordinary least squares (OLS), 73
os module, Python, 180
outliers, observations as, 24

P

p-values
   as basis for decisions, 50-52, 55, 72
   in Excel, 72
   methodology for, 50, 69, 72
   misinterpretations and limitations of, 50, 54, 74
   in Python, 207
   in R, 151, 156
packages
   Python, 77, 161, 171-172, 180, 203
   Python installation of, 171-172
   R, 77, 103-104, 107, 109, 145

R installation and calling of, 103, 107, 145
   updating, 104, 172
pairplot() function, Python, 208
pairplots, 152, 208
pairs() function, R, 152
pandas, Python
   about, 179
   data manipulation with, 187, 192, 193-195
   DataFrames, 179-180, 181, 183, 184
   dtypes in, 177
   frequency tables with, 204
   functions in, 192
   reading in external data with, 181
Pearson correlation coefficient, 62, 65
PEP 8 style guide, Python, 169
Pérez, Fernando, 162
pip Python package installer, 172
pipe (%>%), R operator, 132-133
PivotTables, Excel
   for comparing variables, 19, 22-24
   for data manipulation, 123, 129, 134
   for frequency tables, 12-14
   for t-tests, 48
pivot_longer() function, R, 135
pivot_table() method, Python, 194
pivot_wider() function, R, 135, 146
plotting charts (see visualization of data)
PMF (probability mass function), 35-36
point estimate, 53, 73
population mean, 42, 46, 53
population versus sample data, 19, 41-43, 44, 46, 51, 53, 73, 76
populations in hypothesis testing, 43, 44, 46, 51, 53-55
positive correlation of variables, 62
Power BI, 86
Power Pivot, Excel, 84, 86, 86
Power Query, Excel, 84, 86
Power View, Excel, 84, 86
predict() function, R and Python, 156, 210
predictive systems, 81
print, Python, 169, 180
probability, 27-40
   Bayes' rule in, 28
   central limit theorem (CLT) in, 37-40
   continuous distributions, 32-40
   cumulative distributions, 29, 35
   discrete uniform distributions, 29-32, 37
   empirical (68–95–99.7) rule in, 34-37, 51

with Excel, 65-66, 70, 73
linear/nonlinear relationships in, 62, 65, 70
with Python, 200, 207, 209
with R, 142, 152, 154
regression residuals in, 73
scipy.stats submodule, Python, 203, 207, 208
Script editor, RStudio, 94-96
scripts, R and Python, 96, 162, 165
seaborn package, Python
data visualization with, 187, 195-200, 208, 209
datasets in, 180
select() function, R, 124-126
Series data structure, Python, 179, 188, 189, 191, 207, 208
set.seed() function, R, 155
setwd() function, R, 115
shape attribute, Python, 191
sickit-learn package, Python, 203
simulations for experimental probability, 30-32
sklearn package, Python, 203, 210
slicing in Python, 179, 210
slope in linear regression, 68-69, 70-74
sorting data, 127, 132, 190
sort_values() method, Python, 190
splitting data, 155-156, 209-211
spreadsheets for data analytics, 82-85
SQL (Structured Query Language), 86
sqrt() function, R and Python, 97, 170
squared residuals, 73
stack of tools for data analysis, 81, 88, 213
standard deviations, 18, 34-37, 51, 53, 157
standard error, 52, 53, 72, 75
standard normal distribution, 51
Star dataset example
in Excel, 3, 4-6, 9, 11-26
in Python, 187
in R, 112, 118, 123, 129, 134
variables in, 5-7, 9-11
statistical bias, 43
statistical hypotheses, 42, 44, 54
statistical models, 68
statistical significance, 46-47, 49, 50, 54, 55, 56, 70, 72
statistical testing and methods, 214
(see also hypothesis testing)
statistics
about, 79, 81
descriptive (see descriptive statistics)

further reading on, 147, 214
inferential (see inferential statistics)
statsmodels module, Python, 209, 210
std() function, Python, 192
str() function, R, 102, 110, 114
string data type, Python, 169
Structured Query Language (SQL), 86
style guides, programming, 101, 169
subpopulations, 44, 48, 50
subsetting
arrays, 179
data frames, R, 121
DataFrames, Python, 204
with train/test splits, 155-156, 210
vectors, 112
substantive significance, 51, 54, 55, 57
sum() function, R and Python, 192
summarize() function, R, 124, 130, 134
summarizing variables (see descriptive statistics)
summary() function, R, 119, 153

T

t Stat, Excel, 72
t-distribution, 51
t-Test: Two-Sample Assuming Unequal Variances, Excel, 48-50, 53
t-tests (see independent samples t-tests)
t.test() function, R, 151
tables, creating in Excel, 4
tables, database, 85, 131
tabular data structures, 124, 161, 176, 179
target population, 43
Terminal, launching (Mac), 163, 171
test statistic, 52
testing and training datasets, 155-158, 209-211
testing() function, R, 155
theoretical probability, 30, 31, 33
tibble, R, 117
tidy() function, R, 156
tidymodels package, R, 145, 155-158
tidyr package, R, 135
tidyverse packages, R, 145
versus base R, 115
dplyr (see dplyr, R)
forcats for factors, 114
ggplot2, 136-142
installation of, 103, 109
readr, 116

tilde (~), R operator, 149, 151
toolbar, Jupyter, 165
ToolPak (see Data Analysis ToolPak, Excel)
train/test split and validation, 155-156, 209-211
training and testing datasets, 155-158, 209-211
training() function, R, 155
train_test_split() function, Python, 209
ttest_ind() function, Python, 207
Tukey, John, 3, 93
two-dimensional data structures, 112, 179
two-tailed tests, 47, 50, 51, 54, 69
two-way frequency tables, 13-14, 62, 146, 204
type() function, Python, 169

U

unconditional probability, 28
uniform probability distribution, 29
unique() method, Python, 194
univariate analysis, 62
univariate linear regression, 75
updateR() function, R, 104
updating packages in R and Python, 104, 172
upper() method, Python, 170
usecols argument, Python, 203

V

values argument, Python, 194
value_name, Python, 193
value_vars, Python, 193
van Rossum, Guido, 161
variability, measuring, 17-19, 26
variables
    in advanced analytics, 9
    as columns in data analytics, 5-11
    categorical (see categorical variables)
    classifying, 5-11
    continuous (see continuous variables)
    dependent (see dependent variables)
    descriptive statistics of, 11, 15-21, 26
    discrete, 6, 9, 10-11, 64
    further reading on, 7
    independent (see independent variables)
    linear relationships of, 62-69, 74-76, 208
    versus objects, 100
    quantitative, 8, 10
    storing as factors, 114
    types of, 6, 11, 45
variance, analysis of (ANOVA), 71, 150
variance, measuring, 17-19, 48, 62

var_name, Python, 193
VBA (Visual Basic for Applications), 83
vectors
    defined, 110
    indexing and subsetting, 111
    R data frames as list of, 113, 121, 189
    R data structures with, 110-112, 125, 177
vehicle mileage (mpg) dataset example, 63-66, 69-75, 145, 155, 203
version control systems, 87, 214
View() function, R, 118
Visual Basic for Applications (VBA), 83
visualization of data
    boxplots (see boxplots)
    Clustered Columns, Excel, 14, 23, 29, 32
    countplots (bar charts), 14, 195
    distributions with, 21-26, 148-150, 205
    with Excel, 14, 21-26, 29-32, 34, 36-39, 48, 65, 70, 73
    facet plots, 149, 206
    frequency tables, 14
    further reading on, 13
    with ggplot2, 136-142, 151, 154
    histograms (see histograms)
    line chart, 39
    pairplots, 152, 208
    of probability distributions, 29-40, 32-33, 36-38, 39
    with Python, 195-200, 205-206, 207-208, 209
    with R, 97, 136-142, 148-150, 151-155
    relationships with, 70, 73, 151-155, 207-208
    scatterplots (see scatterplots)
VLOOKUP() function, Excel, 85, 123, 131-132, 193

W

what-if analyses, 56-57
whiskers in boxplots, 24
whitespace in Python, 168
Wickham, Hadley, 136
Wilkinson, Leland, 136
working directories, 115, 122, 181
writexl package, R, 109, 122
write_csv() and write_xlsx() functions in R and Python, 121, 184
writing dataframes, 121, 184

X

x argument, Python, 195, 208
x-axis, mapping, 21, 36, 62, 65, 140-142, 195, 198
.xlsx files, 115, 181, 184

Y

y argument, Python, 197, 208

y-axis, mapping, 62, 65, 140-142, 197, 200

Z

zero correlation of variables, 63
zero-based indexing, 178-179

About the Author

George Mount is the founder and CEO of Stringfest Analytics, a consulting firm specializing in analytics education. He has worked with leading bootcamps, learning platforms, and practice organizations to help individuals excel at analytics. He speaks regularly on the topic and blogs about it at *stringfestanalytics.com*.

George holds a bachelor's degree in economics from Hillsdale College and master's degrees in finance and information systems from Case Western Reserve University. He resides in Cleveland, Ohio.

Colophon

The bird on the cover of *Advancing into Analytics* is a Clark's nutcracker (*Nucifraga columbiana*). This bird, also known as Clark's crow or the woodpecker crow, can be found near treelines on windy peaks in the western United States and parts of western Canada.

The body of the Clark's nutcracker is gray, while its wing and tail feathers are black and white. Its long cone-shaped bill, legs, and feet are also black, and it can reach an average length of 11.3 inches (28.8 cm). The Clark's nutcracker uses its long, "daggerlike" bill to tear into pinecones to extract their large seeds, which they then bury in forest caches for sustenance in the winter. Although these birds remember the majority of their cache locations, what seeds they don't retrieve play an important role in growing new pine forests. The Clark's nutcracker may stash as many as 30,000 seeds in a single season.

The remainder of the diet of the Clark's nutcracker is composed of other seeds, berries, insets, snails, carrion, and the eggs and young of other birds. In part due to their seed stashes, these birds begin their breeding activity in the late winter, nesting on the horizontal limbs of coniferous trees. Both parents care for their young, who typically leave the nest 18-21 days after hatching.

The conservation status of the Clark's nutcracker is "Least concern," although there is evidence that climate change may impact this bird's range and population in the future. Many of the animals on O'Reilly covers are endangered; all of them are important to the world.

The cover illustration is by Karen Montgomery, based on a black and white engraving from *Wood's Illustrated Natural History*. The cover fonts are Gilroy Semibold and Guardian Sans. The text font is Adobe Minion Pro; the heading font is Adobe Myriad Condensed; and the code font is Dalton Maag's Ubuntu Mono.

Milton Keynes UK
Ingram Content Group UK Ltd.
UKHW052011300824
447620UK00007B/211